THE CHEMISTRY OF MEDICAL
AND DENTAL MATERIALS

RSC Materials Monographs

Series Editor: J. A. Connor, *Department of Chemistry, University of Kent, Canterbury, UK*

Advisory Panel: G. C. Allen (*Bristol, UK*), D. J. Cole-Hamilton (*St Andrews, UK*), W. J. Feast (*Durham, UK*) P. Hodge (*Manchester, UK*), M. Ichikawa (*Sapporo, Japan*), B. F. G. Johnson (*Cambridge, UK*), G. A. Ozin (*Toronto, Canada*), W. S. Rees (*Georgia, USA*)

The chemistry of materials will be the central theme of this Series which aims to assist graduates and others in the course of their work. The coverage will be wide-ranging, encompassing both established and new, developing areas. Although focusing on the chemistry of materials, the monographs will not be restricted to this aspect alone.

Polymer Electrolytes
by Fiona M. Gray, *School of Chemistry, University of St. Andrews, UK*

Flat Panel Displays: Advanced Organic Materials
by S. M. Kelly, *Department of Chemistry, University of Hull, UK*

The Chemistry of Medical and Dental Materials
by John W. Nicholson, *School of Chemical and Life Sciences, University of Greenwich, UK*

How to obtain future titles on publication

A standing order plan is available for this series. A standing order will bring delivery of each new volume immediately on publication. For further information, please write to:

Sales and Customer Care
Royal Society of Chemistry
Thomas Graham House
Science Park
Milton Road
Cambridge
CB4 0WF
UK
Telephone: +44(0) 1223 432360
E-mail: sales@rsc.org

**RSC
MATERIALS
MONOGRAPHS**

The Chemistry of Medical and Dental Materials

John W. Nicholson
School of Chemical and Life Sciences,
University of Greenwich, UK

RS•C
ROYAL SOCIETY OF CHEMISTRY

NORTHWEST MISSOURI STATE
UNIVERSITY LIBRARY,
MARYVILLE, MO 64468

ISBN 0-85404-572-4

A catalogue record for this book is available from the British Library

Published by The Royal Society of Chemistry,
Thomas Graham House, Science Park, Milton Road, Cambridge CB4 0WF, UK

For further information see our web site at www.rsc.org

Typeset by Computape (Pickering) Ltd, Pickering, North Yorkshire, UK
Printed by MPG Books Ltd, Bodmin, Cornwall

Preface

The application of materials science to medicine (so-called 'biomaterials science') is a subject of growing importance at the start of the 21st century. Over the last forty years or so, numerous artificial materials (metals, ceramics and polymers) have been used, often in the form of implantable devices, to repair a variety of diseased or traumatised parts of the body. Applications range from orthopaedics (hips, knees, fingers), craniofacial reconstruction, cardiovascular surgery (heart valves, stents, vascular implants), ophthalmic surgery, dentistry and so on.

Somehow, though, implants have not made the same impact on the public mind as transplants, yet the latter suffer from several serious problems. These include the need for extremely high levels of surgical skill, a ready supply of donor organs, and the necessity for constant medication for the patient in order to prevent post-operative rejection of the transplanted organ. By contrast, implants are less demanding of surgical skill, and they require neither a ready supply of donors nor life-long medication for their recipients. Despite their relative neglect by the public, these advantages mean that implants have made a major impact on the health of millions of patients throughout the world. Moreover, with an ageing population, the need for new and improved implantable materials continues to increase.

The development of implants and devices is a multidisciplinary subject, drawing for its success on contributions from *inter alia* engineering, materials science, cell biology and surgery. However, the major emphasis of this book is on the chemistry of these materials, and the book is written from the perspective that chemistry has a significant contribution to make to the subject. Chemistry is important because it tells us about the composition of the materials, the nature of their surface behaviour, their potential for degradation *in vivo* and so on, and all of these aspects impinge on the durability and useful service life of the material or device. There is also the fact that we have a growing understanding of biology at the molecular level – the level at which biology shades into chemistry – so that an understanding of chemistry actually underpins all that we know about the biological interactions of synthetic materials. These include the deposition of proteins on their surfaces, the sequential displacement of these proteins, and later the adhesion of cells to the materials and the biochemistry of their continued viability.

This book is divided into seven chapters. The first deals with the various end uses of implants and devices, and shows how the use of these materials has permeated various branches of surgery. Chapters 2 to 4 deal with the materials used under the conventional divisions respectively of polymers, ceramics and metals. Chapter 5 covers dental materials, and these are dealt with separately for two reasons. Firstly, they are often elaborate hybrids, and cannot easily be classified into one of the major materials groups. Secondly, this is an especially vibrant part of the subject, and many materials have begun life in a dental application, and then found uses in parts of the body well removed from the mouth. Chapter 6 deals with the variety of biological responses, and also safety testing, a critical part of the subject. Finally, the important subject of tissue engineering is touched on briefly in Chapter 7. This is a very large topic and one that includes, but extends beyond, the use of artificial materials in intimate contact with cells, and employed for long-term repair of some part of the body.

Writing a book of this sort, with over 1000 references, is a form of madness. It has, though, also been a labour of love, and certainly made easier by some significant help I have received. Most notably, I was assisted at the planning stage by my friend and former colleague Dr Mary Anstice, now of 3M-ESPE. Circumstances prevented her from continuing as co-author, but I must thank her for her input to shaping the book. I know it would have been a better book had she been able to help write it.

I also want to acknowledge the support of my wife Suzette. No major work I have ever undertaken, beginning with my PhD thesis and extending through all my books, including this one, would have been possible without her patience and forbearance. Once again, I thank her for all her practical help and kindness over the considerable time that I have been engaged in writing this book.

Dr John Nicholson
October 2001

Contents

Glossary

Arthroplasty: The term for the surgical procedure of replacing joints.

Embolism: A blockage of an artery in the lungs by fat, air, tumour tissue or blood.

Fracture mechanics: An approach to the understanding of strength and mechanical failure in materials that is based on a consideration of the influence of cracks (size, geometry and rate of growth) in the material, including at the surface.

Fracture toughness: When a material breaks, due to the extension of pre-existing cracks, two factors influence the amount of energy absorbed. These are (i) the energy needed to create the new surfaces and (ii) the energy absorbed, for example in polymeric materials, by viscous flow of the molecules at the crack tip. The latter causes the material to be tough.

For a crack of original length $2c$, the breaking stress, σ_b is given by the expression:

$$\sigma_b = \left(\frac{K_1}{\pi c}\right)^{\frac{1}{2}}$$

The term K_1 is known as the *fracture toughness* and is a measure of the resistance to growth of pre-existing cracks.

Hank's Balanced salt solution: A solution with the following composition (mmol dm^{-3}):

NaCl	130
KH$_2$PO$_4$	0.44
MgSO$_4$.7H$_2$O	0.811
CaCl$_2$	1.26

NaHCO$_3$	4.1
NaHPO$_4$	0.33
KCl	5.3
Glucose	5.5

pH = 7.4

Ringer's solution: A solution with the following composition (mmol dm^{-3}):

NaCl	137
KCl	5.64
CaCl$_2$	2.16
NaHCO$_3$	2.38

pH = 7.2

It is isotonic with cells and is used for studies on mammalian tissues. The term is sometimes loosely applied to any salt solution that approximates to physiological saline.

Stem cells: Cells that have the ability to divide for indefinite periods in culture and to give specialised cells.

Strength: The maximum load a material or structure can bear before failing. Depending on the orientation of the load, strength can be classified as:

(i) tensile (the loads are applied in tension, *i.e.* pulling the structure apart);
(ii) compressive (load is applied to crush the specimen);
(iii) flexure (load is applied in such a way that the specimen breaks by bending).

Other specialised types of strength are often determined for biomedical materials, notably (a) biaxial flexure, where a disc of material is supported on a ring and broken by loading in the centre; and (b) diametral tensile strength, where a cylindrical specimen is loaded along its axis, resulting in forces along that axis that are essentially tensile in nature.

Wear: The loss of material that comes as a result of rubbing at the surface. It is usually classified as two-body (two surfaces rubbing against each other causing material loss) or three-body (two surfaces, plus an abrasive powder, rubbing together).

CHAPTER 1

Synthetic Materials in Medicine

1 Introduction

The use of synthetic materials in the body by medical and dental practitioners
to provide repair and function has grown remarkably in the last 30–40 years,
though the concept of using such artificial materials is very old. For example,
Plaster of Paris was pioneered as bone-substitute material towards the end of
the nineteenth century, and dental fillings, including amalgam, have been
around for well over 150 years. The use of engineered structures fabricated
from metals and polymers in orthopaedic surgery has a more recent history,
however, beginning with Dr (later Sir) John Charnley's work on the replacement of arthritic hips in the early 1960s.[1,2] This surgical repair technique,
known as total hip arthroplasty, has seen particularly spectacular growth, and
since Charnley's original cemented hip replacements there have been a variety
of new materials and new designs for implantable devices, and these are now
available not only for hips, but also for knees, toes and fingers.

Synthetic materials used in the body in this way are widely referred to as
biomaterials. This use of the term appears to have originated in 1967 with the
first 'International Biomaterials Symposium' at Clemson University, South
Carolina, since which time it has been used extensively in this way. In many
ways to apply the word *biomaterials* to synthetic materials is not very
satisfactory since by analogy with, for example, the word biochemistry, it
might be assumed to refer to materials of biological origin. However, within
the field of implantable devices, the word *biomaterial* has been formally defined
as *a non-viable material used in a biomedical device intended to interact with
biological systems*.[3] This definition was adopted at the Consensus Conference
of the European Society for Biomaterials, held at Chester, UK, in March 1987,
and has been widely accepted ever since. In fact, some sort of definition of this
type was already implicit in the title of the organisation which ran the meeting,
the European Society for Biomaterials, because the Society's object from the
time it was established in 1976 was to promote the study the science of such
synthetic materials. It was never primarily concerned with the science of
natural substances, such as teeth or bones. The current definition was also

1

implicit in the title of the scientific journal *Biomaterials*, which was first published in 1980. Whatever the rights and wrongs of the etymology, by usage the term *biomaterial* has now clearly come to mean a synthetic material with a biological destination rather than a biological origin.

There is a further caveat with the term, in that it is usually applied to materials designed to reside within the body for some considerable time. Thus, materials used to fabricate devices used only in surgery, ranging in sophistication from sensors to catheters, are not usually regarded as biomaterials. They may interact with a biological system, the body, but such interaction is usually relatively brief. Sutures, too, are not usually regarded as biomaterials for a similar reason. On the other hand, degradable polymers of the type used in sutures are finding increasingly novel uses in medicine, for example as temporary scaffolds and supports for bone immobilisation. These enable the body's own repair mechanisms time to bring about complete healing without premature loading and potential failure and under these circumstances, the polymers become biomaterials, because their interaction with the body must continue for a considerable time.

The field of biomaterials science encompasses all classes of material, *i.e.* polymers, ceramics, glasses and metals, and a wide range of branches of surgery: dental, ophthalmic, orthopaedic, cardiovascular and so on. The key requirement of any material or combination of materials used in the body is that, in addition to providing mechanical support or repair, it should be *biocompatible*. The subject of biocompatibility is covered in detail in Chapter 6, but at this stage we should note its definition. This is *the ability of a material to perform with appropriate host response in a specific application.*[1] As stated in this definition, biocompatibility is not a property of a material *per se*; the material needs to elicit an appropriate response, and whether such a response is appropriate will depend on the site in the body at which it has been placed. A material which shows excellent biocompatibility, for example, in contact with bone would not necessarily show good biocompatibility when used in a blood-contacting device, such as an artificial heart valve. Thus the location within the body is as important in determining whether a material is biocompatible as the composition of the material.

The property of biocompatibility is distinct from that of inertness, which would imply a complete absence of response from the body. At one stage, it was thought that inertness was a desirable property, but nowadays inertness is not thought possible. Even materials which seem inert in most technical applications, such as polytetrafluoroethylene, PTFE, prove to be highly active when placed within the body. PTFE was once used to fabricate the acetabular cups used in experimental hip replacement surgery.[4] When used in conjunction with a metal femoral head, it proved to have extremely poor wear characteristics, leading to build-up of high local concentrations of particulate wear debris. This wear debris provoked extreme adverse reactions in patients, leading to severe swelling and general discomfort.[5] Consequently, the use of PTFE for this purpose was abandoned.

Because of experiences of this type, there has been a shift in thinking and the emphasis nowadays is on materials that will elicit a response from the body that is appropriate.[6] This may be, as in the case of titanium implants, anchorage without formation of fibrous capsule.[7] Although it desplays this desirable feature, titanium is by no means inert the human body. It may undergo corrosion,[8] and this can be so severe that the tissues close to the implant become darkened by the build up of titanium within the cells.[9] Despite the potential for such adverse effects, in general the presence of titanium is well tolerated by the body,[10] and the use of titanium for the fabrication of implants is a current feature of many branches of surgery.[11]

The successful use of biomaterials presents numerous challenges. A major one is the issue of maintenance, and in particular that most devices are implanted well into the body and therefore not freely available for inspection or repair. An artificial hip joint, for example, is completely inaccessible, except by major surgery, and hence cannot be routinely serviced. The body is a hostile environment, despite its sensitivity, and it provides very severe service conditions. In no other field of technology are manufactured items expected to function without maintenance for so long in comparably demanding conditions.

Life expectancy in the wealthier parts of the world is now of the order of 80 years, which means that many people now outlive the useful life of their own connective tissue.[12] This can be be quantified by considering what happens to the bones on ageing. In men, between the ages of 20 and 60, there is a 25% loss of cortical bone; in women, this loss is 35%.[13,14] These changes lead to losses in strength of the order of 80–90%,[15] and increases in the risk of bone fracture over the age of 50 of 13% in men and 40% in women.[16] As a consequence, by 80–90 years of age, 33% of women and 17% of men will have hip fractures.[17]

When synthetic materials are used to effect repairs, they must be able to survive for considerable lengths of time and without maintenance. However, it is rare to find an implant whose life expectancy exceeds 15 years, regardless of whether that implant is designed for orthopaedic, cardiovascular, dental or other application. This represents the major challenge in the field, and one that is extremely elusive. Despite much research activity in biomaterials science, the problem of maintenance-free durability remains with us, and there have been only marginal extensions in the anticipated lifetimes of implants as a result of considerable volumes of research in both materials and surgical techniques. On the other hand, what has been achieved is remarkable, and there is no doubt that biomaterials alleviate suffering and add to the quality of life for a very large number of individuals throughout the world.

2 Surgical Uses of Biomaterials

Typically, biomaterials are fabricated into a medical device of some sort, and employed in the body in this form. In this context, the term *device* has been defined by the United States Federal Drug and Food Administration (FDA) as '. . . any instrument, apparatus, implement, machine, contrivance, *in vitro* reagent or combination of these that is intended for diagnosis, prevention or treatment of disease'.[18] This is a comprehensive definition, and includes those applications of synthetic materials in the human body where the material itself is simply placed, perhaps taking up the shape of a specially prepared cavity, as in dental filling materials or orthopaedic bone cements. Devices so defined are currently employed in a wide range of branches of surgery to treat a variety of conditions. Some of the more frequently employed examples of these are described in detail in the remaining sections of this chapter.

Orthopaedic Joint Replacement

As the body ages, it becomes susceptible to osteoarthritis, a condition in which the lubricating layer of cartilage covering the bone in joints degenerates.[19] This degeneration results in loss of freedom of movement of the joints, together with extreme pain. Both features contribute to the immobilisation of the patient, with potentially serious secondary effects on the health resulting from reduced exercise. The main joints that are affected are the hip and the knee, and treatment is by joint replacement, a procedure known as total joint arthroplasty. In this operation, the diseased joint is removed and replaced with one composed entirely of artificial materials. These maintain their location relative to the bone, and are required to survive preferably for the remaining lifetime of the patient.[20] Hip replacement is the most widely performed of the total joint arthroplasties, and an estimated 287000 are performed each year throughout the world.[13]

The hip and knee, and also other joints, such as the shoulder, are complex and delicate structures. They comprise an optimised combination of articular cartilage, bone and synovial fluid.[21] Articular cartilage is a connective tissue that covers the bones, and is capable of bearing loads. It has a low coefficient of friction, thereby contributing to the natural lubrication of the joint.[22] Synovial fluid is a nutrient solution secreted within the joint,[11] which also partly exhibits a lubricating function. Human joints are prone to degenerative diseases, of which osteoarthritis (loss of articular cartilage) is the major one, but others also include rheumatoid arthritis (swelling of the synovial membrane) and chondromalacia (softening of the cartilage). Approximately 90% of the population over the age of 40 suffers to some extent from degenerative bone disease, a state of affairs which arises either because of excessive loading of the joints during the early life of the patient or because of failure of the normal repair processes later in life leading to degeneration. Such failure occurs by a mechanism or mechanisms which are not currently understood.

In total hip arthroplasty, the natural hip joint is replaced by an artificial one consisting of a femoral component that is usually a polished metal ball mounted on a metal stem, and an acetabular component that has a socket in which the ball sits and swivels as the patient walks. The femoral component is predominantly metal, either cobalt-chrome or a titanium alloy; the acetabular component, sometimes called a cup, is made from ultrahigh molecular weight polyethylene, UHMWPE. The use of a head smaller than that of the natural femur was one of the pioneering developments of Charnley. A small head makes the pin easier to place and finish than a larger head, though the latter approach, using a head size equivalent to that of the natural femur, was used for some time in the so-called McKee–Farrar hip. The components of a Charnley-type hip are shown in Figure 1.1.

There are various designs of femoral pin and head in use. Stems vary in length, curvature and finish, and also in whether or not they have collars or other mechanical additions for spreading the load or integrating with the bone.

The components of the hip are generally held in place with a self-curing acrylic bone cement. This was originally based on the same material as denture bases, but has since undergone considerable study and modification. The cement is generally prepared by mixing methyl methacrylate monomer with pre-polymerised beads of poly(methyl methacrylate) that also contains a polymerisation initiatitor, typically benzoyl peroxide with dimethyl-*p*-toluidine as accelerator. On mixing, the monomer undergoes a rapid polymerisation and sets to yield a reasonably strong and rigid material within 10–20 minutes of mixing. The detailed chemistry of such bone cements is covered in Chapter 2. It been claimed that the use of cement was the major factor in the success of this surgical technique, and prostheses of this type, cemented by poly(methyl methacrylate) have a success rate of 85–95% at 15 years.

Early hip replacements were often complicated because of a high incidence of infection. These problems were overcome by improvements in operating room procedures and the use of antibiotics following surgery.[23] There remain failures with the technique, of which the most common is so-called 'aseptic loosening', in which the femoral component of the joint becomes loose due to bone resorption around the cement. This has been assumed to be caused *inter alia* by individual patient intolerance of the poly(methyl methacrylate) bone cement and attempts have been made to overcome the problem by using a cementless surgical technique. This has often been accompanied by the use of specially prepared metal prostheses, with either roughened surfaces or with surfaces coated in plasma-sprayed hydroxyapatite.[24] Both of these surface-treatments are aimed at encouraging the growth of bone right up against the implant, a process known as *osseointegration*. Despite initially encouraging results, currently operations performed with cemented prostheses have superior outcomes, both in terms of patient mobility following the operation and survival rates of the implant.

A general problem with all artificial joints is that of stress transfer from synthetic materials, whose moduli are typically very different from that of bone.[25] In most cases, following the insertion of an artificial hip, the outer wall of the femoral cavity becomes more heavily loaded than in the natural hip, whereas the inner wall becomes more lightly loaded. Bone is a dynamic tissue whose form is maintained by a balance between the activities of the bone depositing cells (osteoblasts) and the bone resorbing cells (osteoclasts).[26] Bone is a piezoelectric material,[27] and thus responds when its pattern of loading changes. This alters the relationship between the activities of the osteoblasts and the osteoclasts. Following hip replacement, this change leads to the deposition of extra bone along the outer wall of the femur and the removal of

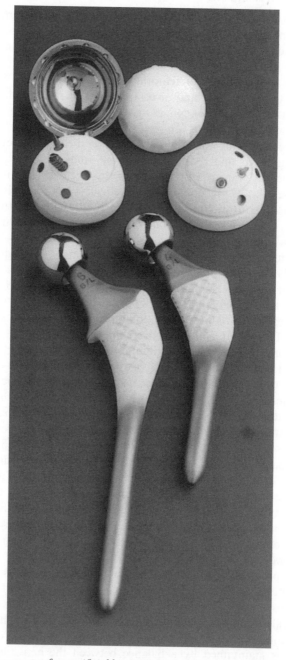

Figure 1.1 *Components of an artificial hip-joint for use without bone cement* (courtesy of Howmedica-Stryker Ltd)

bone from the inner wall. The inner wall is often described as experiencing 'stress shielding' and in extreme cases this can lead to the development of holes in the femoral wall, with disastrous effects on the overall strength of the replaced hip joint.

A final difficulty to be considered is that of wear of the acetabular cup by the artificial femoral head.[28] In recent years it has been realised that such wear is a major problem and it is currently receiving considerable attention from researchers. One approach to deal with it is to use ceramic femoral heads. Ceramics, typically alumina or, less commonly, zirconia, cause different wear behaviour of the UHMWPE acetabular cap than metals.[29] If a metal is scratched, it tends to creep in ductile fashion, lifting sharp edges along the length of the scratch, and these scrape out pieces of the polyethylene. By contrast, if a ceramic is scratched, because it is a brittle material, it tends to respond by losing material from its surface. This behaviour results in the retention of a smoother finish than the metal, and one which is still able to articulate with the polyethylene without scraping out debris. Because of these differences, ceramic heads cause considerably less damage to acetabular cups than metal ones, with correspondingly fewer problems of wear debris and adverse tissue response due to the accumulation of polymer particles in the region around the joint.[30]

After hips, the next most commonly replaced joint is the knee. Total knee replacement surgery was developed somewhat later than hip replacement, but since about 1974 has become a successful surgical technique. Knee replacement is required for similar reasons to hip replacement, most frequently degenerative loss of lubricating cartilage within the joint, leading to severe pain and progressive immobility. Replacement of the knee is more complicated than replacement of the hip, due to its more complex pattern of loading.[14] This arises partly from the concentration of force through the knee when an individual is standing, and partly because of the rotational motion of the tibia relative to the femur during walking.[31]

The components of the natural knee joint are illustrated in Figure 1.2. To replace this with an artificial joint requires a number of features.[32] The articulating end of the femur needs to be replaced by a lightweight component that typically consists of a thin, rigid shell with an attached fixation system to anchor the device to the bone. Details of the construction vary from device to device, and may include a shaft to stabilise the component relative to the femur. Metals, either a cobalt-chrome or a titanium alloy, are used to fabricate this component, since they combine the necessary properties of high strength, high modulus and low wear rate. This component is either cemented in place with a poly(methyl methacrylate) bone cement, or finished in such a way that full osseointegration can occur readily.

To replace the tibial aspect of the joint, a relatively broad platform is inserted. This typically consists of a stiff metal tray, with a polymeric component, usually made from UHMWPE, providing the articulating surface. This UHMWPE component is subject to very high loading as the patient walks, and this leads, on occasion, to failures due to creep or fatigue. As with

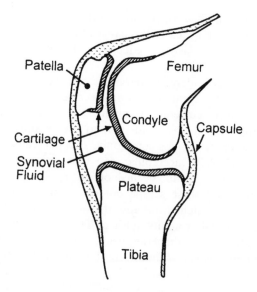

Figure 1.2 *Diagram of the natural knee joint*

hip joints, the UHMWPE can experience significant wear due to the repeated motion of the metallic component in contact with it.[23] The topic of wear of UHMWPE and its biological effects is dealt with in Chapter 2.

The bone into which these replacement joints is placed has a complex structure. It consists of hydroxyapatite together with the fibrous protein collagen and water, as well as minor components, some of which are critical for bone to perform its biological function. The mineralised tissue of vertebrates is different from that of other living groups in that it represents a more extensive component of their bodies and also that the mineral phase is a form of calcium phosphate, rather than calcium carbonate or silica.[33]

There have been numerous studies of the mechanical properties of bone, but the results are difficult to interpret. This is because bone not only behaves like all materials in giving different values of strength depending on loading regime, *e.g.* whether in tension, compression or flexion and on rate of loading, but also it is anisotropic, *i.e.* its properties are not the same in all directions.[28] Properties also change with age and with the health of the individual.[34] For example, in a study of the human femur, the tensile strength was found to vary from about 120 MPa at age 20 to 65 MPa at age 95.[35] Similarly ultimate tensile strain went down from 3.5% to about 1% over the same age range. By contrast the tibia showed much less change.[36]

The organisation of bone varies with bone type. Broadly speaking, the structure involves the interaction of the two solid phases, collagen and hydroxyapatite. Hydroxyapatite crystals occupy gaps between the collagen fibres[37] and are bonded to the collagen through interactions of the polar groups on the protein molecules with the calcium phosphate crystal structure.[38] The organisation of the collagen varies between the types of bone. In woven bone, which is laid down rapidly, typically in the foetus or in the callus that forms during the repair of fractures,[28] the collagen fibres show little orientation. In lamellar bone, the collagen is more ordered, with collagen fibrils in individual lamellae being lined up in more or less the same direction, but at right angles to the fibrils in the immediately adjacent lamellae.[39] Parallel-fibred bone also occurs, and in this the collagen fibrils are all aligned with the long axis of the bone.[40]

Tendons and Ligaments

Tendons and ligaments are tough, collagenous tissues with connective functions. Tendons connect bone to muscle, while ligaments connect bone (or cartilage) to bone. They are flexible and capable of contraction, and because of these properties, are able to move the bone or cartilage to which they are attached.

There are at least twelve different types of collagen found in nature, of which types I, II and III predominate in mammalian tissues.[26] These types form the structural fibres in tendon and also skin, cartilage and cardiovascular tissue. They are triple-helical molecules, mainly based on the amino acids glycine, proline and hydroxyproline. Tendon also contains type V collagen, whereas ligament is made up of type IV collagen.[26]

Tendon and ligament damage is most often sustained through injury to athletes engaged in contact sports, but may also occur in military personnel, labourers and factory workers. In cases of sports injuries, transplanting of autogenous tissue has been a successful technique for restoring normal joint function, for example of the knee.[41,42] The problem with this approach is that recovery time following surgery is prolonged. Alternatively, where a tendon or ligament is damaged but not completely torn, immobilisation of the joint leads to healing with the formation of scar tissue. This is also not completely satisfactory, because scar tissue is inferior biomechanically to undamaged tendon or ligament; it is weaker and less readily contracted.[43]

The most frequently damaged ligament is the anterior cruciate ligament of the knee.[14] Injury to this ligament is increasing, due to the increase in athletic activity in many countries, notably the United States. Replacement of this ligament is not straightforward and no medical devices employing synthetic materials have yet been approved by the United States Food and Drugs Administration for this purpose. Because of this, biological graft material is often used as a substitute for the natural ligament. However, various synthetic materials have been studied experimentally and a number of devices have received conditional approval for clinical use. These include carbon fibres, poly(ethylene terephthalate), polytetrafluoroethylene and braided polyethylene.[14] Composite structures, such as carbon fibre coated with poly(lactic acid), have also been examined,[44] though concern has been expressed about the fate of carbon particles generated in use once the poly(lactic acid) has degraded.[14]

There have also been problems reported with the morphology of the collagen induced in the region of artificial ligament materials.[45] Synthetic

Table 1.1 *Materials used in cardiovascular surgery*

Size	Material
12–38 mm diameter	Poly(ethylene terephthalate)
5–10 mm diameter	Poly(ethylene terephthalate), PTFE
< 4 mm diameter	Autografts

polymers, such as UHMWPE, have been shown to possess inadequate fatigue properties for the replacement of the anterior cruciate ligament of the knee.[46] The general conclusion from these findings is that synthetic polymers do not make acceptable permanent substitutes for ligaments. Instead, the use of such synthetic materials should probably be restricted to the fabrication of temporary (degradable) scaffolds. Materials for use in this application are typically poly(glycolic) or poly(lactic acid), or related copolymers, and they degrade in the body over time periods ranging from a few weeks to several months, depending on the precise details of their composition. To date, these materials have been mainly studied experimentally using animal models, and the amount of work in human patients has been very restricted.

Cardiovascular Implants

Diseases of the cardiovascular system contribute to approximately 20% of the deaths of people between the ages of 36 and 74 throughout the world.[14] A major cause of death is the condition known as atherosclerosis, which affects large and medium diameter blood vessels, especially the aorta, coronary arteries and cerebral arteries. In this condition, fatty deposits and fibrous tissue partly formed from collagen are precipitated on the internal wall of the blood vessel, narrowing the diameter. This leads to abnormal dilations of the affected blood vessels known as aneurysms, which can lead to tearing or bursting of the vessel walls, with subsequent death.[47]

Another possible defect of the cardiovascular system is leaky heart valves, which like atherosclerosis become more common with advancing age. Both athereosclerotic blood vessels and leaky heart valves are now routinely replaced, either with synthetic or natural materials. Examples of materials that have been used are given in Table 1.1.

There are over 5000 artificial heart valves implanted annually in the United Kingdom, and over 100 000 in the United States.[48] For such heart valve replacement, two approaches have been used, either mechanical devices involving a disc or ball in a metal framework, or a bioprosthetic valve, fabricated for example from bovine or porcine tissue. The mechanical device is more common, and is used in about 75% of all heart replacement cases.[49] Neither type is fully satisfactory, and each may encounter problems in service. The synthetic heart valve requires patients to remain on anticoagulant medication for the rest of their lives in order to prevent clotting of the blood in the region of the implanted material. On the other hand, though patients receiving

Table 1.2 *The composition of blood*[51]

Substance	Vol%	Comments
Water	35	
Proteins		
Albumin	1–2	Molar mass 69 000
Fibrinogen	0.2	Molar mass 34 000
Ions	<0.5	Molar mass <100
Cells		
Red blood cells	45	Biconcave disc, 8 μm × 1–3 μm thick
White blood cells	<1	Spherical, 7–22 μm diameter
Platelets	<1	Disc, 2–4 μm diameter

bioprosthetic valves do not require permanent anticoagulant treatment, these valves are prone to failure by calcification and degeneration.[14]

As a matter of course, all cardiovascular prostheses must come into contact with blood, and ensuring that the blood shows an appropriate response to the presence of foreign bodies is a major challenge. In particular blood is very prone to clot, and one approach to enhance the haemocompatibility of polymeric materials in particular has been to alter the hydrophobic/hydrophilic nature of the surface. Extremes of either hydrophilic or hydrophobic character seem to give the best results for haemocompatibility. For example, in one study the surface of a polyurethane was modified by grafting either perfluoro-decanoic acid to produce a more hydrophobic surface or poly(ethylene oxide) to yield a more hydrophilic one.[50] Both approaches were found to improve the blood compatibility compared with that of the untreated polyurethane surface as deteremined by the *ex vivo* clotting time of blood in contact with the polymer surface.[45]

The composition of blood is shown in Table 1.2.

The coagulation of blood (clotting) occurs readily, and this is desirable because it defends the body against excessive blood loss when there are small leaks in the cardiovascular system. Clotting is a complex process, and may be initiated either *via* the surface, the so-called intrinsic cascade, or *via* tissue damage, the extrinsic cascade. These lead to the formation of complexes between the blood proteins and the platelet surfaces, followed by activation of prothrombin to form thrombin and fibrinogen to form a fibrin clot. Coagulation of the blood can also trigger the kinin system, which results in leakage of fluid from the vascular system into the surrounding tissues, a process that results in inflammation. Synthetic materials can interact with various stages of the coagulation pathway, thus provoking thrombus formation. They can also cause prolonged inflammation, which may compromise implant function and cause a variety of unacceptable side effects.[14]

The initial interaction between the body and a synthetic material may be influenced by several factors, including composition, surface energy, roughness and topography.[52] A cascade of events has been proposed to describe the immediate effects of implanting a material within the cardiovascular system.[53] This cascade begins with the arrival of water molecules at the implantation site, where they hydrogen bond *via* hydrogen or oxygen atoms. Ions become incorporated into this water layer, and these are followed by the arrival of proteins, which interact with them and with the water molecules to become associated with the layer. Gradually, there is a build-up of biomolecules in this layer, the precise details of which vary depending on the implanted material. However, the nature of the proteins and other biomolecules deposited determines which serum components present in the extracellular fluid actually become adsorbed on the surface. The most common substances adsorbed are fibrinogen, gamma-globulin and albumin, but other substances may also be deposited. The amounts deposited are related to their concentration in the extracellular fluid. Once this plasma protein layer has been adsorbed, a cross-linked fibrin network can develop. If fibrin formation is too rapid, a thrombus may develop. On the other hand, if it occurs slowly, the process of fibrinolysis may be activated, and thrombus formation prevented.[47]

Overall, the goal of developing biomaterials that show full haemocompatibility is probably the toughest challenge facing scientists in this field. The sensitivity of the blood towards coagulation, and the ease with which foreign materials trigger the process, means that implanting devices into the cardiovascular system is always potentially difficult. Although good progress has been made, as evidenced by the widespread use of items such as synthetic heart valves and other devices, there remains a long way to go, and the ideal of a synthetic material that can be implanted without the patient needing to be placed on permanent anticoagulant therapy is still to be realised.

Ophthalmic Implants

Synthetic materials are used to improve and maintain vision, a requirement of growing importance as the population continues to age. The eye itself is a complex organ that is held within the bony orbit of the skull by extrinsic eye muscles. The structure of the eyeball is shown in Figure 1.3.

The eyeball is about 2.5 cm in diameter and is made up of three layers, the fibrous outer coat, the vascular middle layer and the light sensitive inner lining. In the back of the eye is a thin pigmented membrane called the choroid which supports the retina. The retina is a light sensitive part of the eye which functions by converting light intensity and colour into electrical signals, for transmission to the brain *via* the optic nerve. As well as its pigmented epithelium, the retina contains cells known as rods and cones. The rods are sensitive to dull light and are essentially responsible for the perception of movement and shape. The cones are sensitive to bright light and are responsible for the perception of colour and sharp outline.

The eyeball is filled with fluid. The main part, behind the lens, is filled with the vitreous humour, a viscous, gel-like liquid containing thin collagen fibrils. The liquid itself is a solution of hyaluronic acid, which is a linear polymer of D-glucaronic acid and *N*-acetylglucosamine. The molar mass is highly poly-

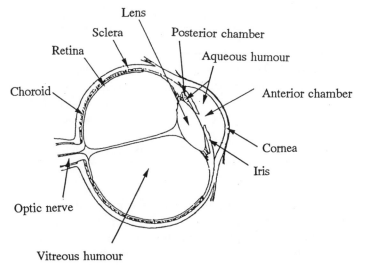

Figure 1.3 *Structure of the eyeball*

Table 1.3 *Materials used in intraocular lenses*

Component	Material
Optical portion	PMMA, poly(2-hydroxyethyl methacrylate), silicone
Stabilisers	Antioxidants, UV absorbers
Anchor	Metals, nylon, glass, polyimide, polyethylene

disperse, and ranges from 77 000 to in excess of 100 000.[54,55] In front of the lens is a chamber containing the aqueous humour, which, as its name suggests, is a low viscosity liquid. It is maintained at a pressure of about 24 mmHg by a balance between continual production and drainage of the fluid.[14]

The lens is made up of collagen, cell attachment factors and proteoglycans characteristic of the lens tissue.[14] It also contains an unusual class of protein called crystallins that are arranged in such a way that they are transparent.[56] A common problem in older patients is loss of transparency in the lens leading to cataract formation. This in turn results in loss of vision.

Treatment of cataracts is now routinely achieved by the implantation of synthetic intraocular lenses. These lenses consist of an optical portion through which light passes, and a loop portion for anchoring the lens within the eye.[57] The optical portion is typically made from poly(methyl methacrylate), PMMA, which has excellent clarity and is very biocompatible when implanted into the eye. However, other materials have also been used, as shown in Table 1.3.

Typically, intraocular lenses are placed in the posterior chamber of the eyeball, though anterior chamber placement is also used occasionally. This latter approach tends to be used if the eyeball has been severely damaged, for

Table 1.4 *Materials used for rigid gas-permeable contact lenses*

Polymer	Comments
Acrylate with microscopic holes	Increased permeability
Fluorinated methacrylate or pyrolidone	Reduced corneal swelling for up to 60 days
Silicones	Hydrophobic, hence interact with insoluble tear components and preservatives in lens solutions
Silicone-methacrylates	Not permeable enough for extended wear

example when the posterior chamber has been ruptured and there has been loss of the vitreous humour.[58] Posterior chamber placement is achieved by removal of the natural opaque lens and insertion of the synthetic lens into the vacant capsular bag. This is not straightforward, and a major difficulty with the use of intraocular lenses is ensuring that they are adequately fixed for long-term service.[59]

Although PMMA is the most widely used lens material, other polymers have been used, notably silicones. These have the advantage of being flexible, so can be folded for insertion through incisions. This causes less damage to the endothelium than the placement of PMMA lenses, a feature that may be important in certain patients.[60,61]

Polymeric materials are also used to fabricate contact lenses as an alternative to spectacles for the correction of defective vision. As for intraocular lenses, poly(methyl methacrylate) is widely used for this purpose. However, it is rigid and also has a low oxygen permeability, both of which are disadvantageous for this application. Delivery of an adequate supply of oxygen to the cornea is vital for the health of the eyeball, and for this reason contact lenses showing higher rates of oxygen diffusion have been developed.[62]

Initially, soft contact lenses were developed because diffusion coefficents for oxygen are higher through these gels than through rigid polymers. However, soft contact lenses become less permeable to oxygen as they age. They are also susceptible to deposition of lipomucoprotein from the tear film and to growth of fungi and bacteria. For this reason, rigid gas-permeable materials have been developed,[42] examples of which are listed in Table 1.4.

Despite these developments, it is still difficult to find a material suitable for extended wear. Unless lenses are removed at least for periods of sleep, and preferably at other times, complications may arise. These include corneal abrasions, hypersensitive reactions, hypoxia, infections and toxicity.[63] Research is still underway to try to solve the problem of extended wear of contact lenses, and there is little doubt that, if successful, such lenses would prove popular.

Defects in vision may not only occur because of problems with the lens. There is also the problem of glaucoma. This is a condition in which the pressure of the intraocular fluid becomes elevated, leading to progressive damage to the optic nerve, with resulting loss of vision. The main surgical treatment is to insert an opening that allows fluid to drain through the sclera/cornea into the subconjunctival space. For some patients, it is advantageous to place a drainage implant at this site to facilitate the process. The first successful device of this type was described by Molteno in 1969, and comprised a two-piece device consisting of a silicone drainage tube and a polypropylene plate.[64] The plate was sutured in place under a flap and had the function of maintaining the patency of the subconjunctval filtration reservoir during subconjunctival fibrosis. Since this time, a number of similar devices have been developed, typically with silicone drainage tubes but with plates fabricated from various polymers, including silicone and polymethyl methacrylate.[65]

Finally, there are a number of other uses for synthetic materials in ophthalmology other than for the lens or glaucoma drainage. These include artificial tear solutions for patients with dry-eye conditions.[66] Such solutions typically contain a variety of substances, such as methylcellulose, poly(vinyl alcohol), and hyaluronic acid, though formulations vary, and there is no ideal formulation to suit the majority of patients who suffer from dry-eye syndrome.[67]

Dentistry

The structure of a typical tooth is shown in Figure 1.4. The destruction of its fabric, dental caries, though preventable, is one of the most widespread diseases of civilisation.[68] Dental caries involves attack on both the enamel and the dentine of the tooth by acids that are generated as the result of metabolic activity of bacteria within the plaque that accumulates on the tooth surface. The principal acid that is formed is lactic acid, but acetic acid is also generated in smaller amounts by active caries.[69] Initial attack is quite localised on the surface of the enamel, but as it penetrates the softer dentine, it tends to balloon out, and undermine the surface layer of enamel.

Diet makes the major contribution to dental decay, especially the presence of refined sugar. Sugars are the food source for bacteria in the plaque, and are metabolised through to the acids that cause the damage. The pH of the mouth drops rapidly after the intake of sugar, as was first demonstrated by Stephan in his classic study.[70] He gave his subjects a mouth rinse of glucose solution, then measured the pH at the surface of the tooth over the next hour. His findings are shown in Figure 1.5.

The critical factor in caries development is the time for which the pH stays below 5.5. Stephan's results showed that in patients who already had a significant amount of tooth decay, the pH stayed below this critical value for longer than it did in those without decay.

Sucrose, and to a lesser extent glucose, is metabolised to form intracellular and extracellular polysaccharides. These sugars enable the bacteria to cling on tenaciously to the tooth surface, and provide a source of energy for continued metabolic activity. The main species responsible for caries is *Streptococcus mutans*, though other bacteria are also involved.

Tooth decay is mainly a problem among the young, who tend to consume

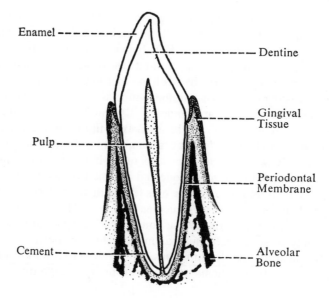

Figure 1.4 *structure of a tooth (incisor)*

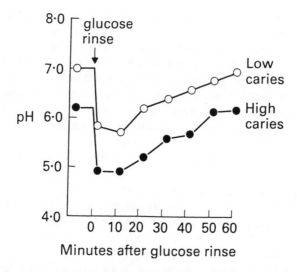

Figure 1.5 *Curve of pH vs time following administration of a glucose drink* (from Stephan[70])

sweets, and also among the elderly, who often suffer from dry-mouth syndrome as a side effect of medication, and therefore resort to consuming sweet drinks and to sucking boiled sweets to relieve the discomfort. These patients are increasingly likely to have at least some of their natural teeth, and these teeth become susceptible to loss from caries as a result of the increase in sugar consumption.

Tooth decay is generally repaired by removing the carious tissue and replacing it with an appropriate restorative. The most widely used and cost effective materials for this purpose is silver amalgam, though there has been considerable work on polymer-based restorative materials in recent years, and this has led to tooth-coloured alternatives to silver amalgam. These are generally bonded to the tooth using either the inherent adhesive nature of the restorative, or special bonding agents.[71]

Tooth loss, which may result from severe caries, or from periodontal disease in middle life, has conventionally been treated by providing dentures, either full or partial, depending on the severity of the loss. More recently, partial tooth loss has been treated with the use of implants.[72]

Dental implants have mainly been used as replacements for tooth roots or tooth root analogues. They have also been used successfully to facilitate orthodontic tooth movement and also for prosthetic treatment of craniofacial defects. In selecting a patient for implant treatment, certain biochemical and biomechanical requirements should be taken into account. For example, there should be no disease that would compromise wound healing. Conditions such as diabetes, osteoporosis and various cardiovascular diseases may cause concern, although they do not necessarily mean that implant treatment cannot be used.[73] In general, though, patients should be in good health, be psychologically stable, and have adequate bone density at the site of the proposed implant. Dental health should be good and especially healthy oral tissues are required.

The main material used for dental implants is titanium and its alloys. These are used because of their excellent bone biocompatibility and ability to osseintegrate.[74] To replace teeth, titanium support structures are implanted into the jaw, and used to support ceramic teeth of excellent aesthetic appearance within the mouth.

As well as titanium-based implants, other materials have been used in dental implantology. For example, synthetic hydroxyapatite in particulate form has been used to improve healing in patients who have had periodontal surgery.[75] When used in this way, the hydroxyapatite particles become surrounded by collagen early in the healing process, and gradually diminish in size. At the same time, bone is gradually deposited in the region around the particles. Synthetic hydroxyapatite of this type may also be used to augment the alveolar ridge in edentulous (*i.e.* toothless) patients, thereby providing improved support for dentures.

Wound Dressings and Artificial Skin

Wounds to the skin in terms of minor cuts and abrasions are common occurences, and usually need no more elaborate treatment than to be kept clean and dry. For more extensive skin damage, for example severe burns, synthetic materials are used, either in the form of sterile dressings or as artificial skin. These limit the entrance of foreign matter into the skin and also prevent loss of fluid and heat from the surface of the skin.

In a typical person, the skin occupies an area of approximately 2 m^2. Skin tissue has a variety of functions, including regulation of body temperature, wound repair, protection from disease and removal of waste.[76] It also serves to protect the body's internal organs from injury.

The skin is divided into two distinct regions, the outer part, called the epidermis, and the inner part, called the dermis.[77] The epidermis varies in thickness, depending on its location, and is thickest where the skin needs to bear the highest loads, *i.e.* on the palms of the hands and the soles of the feet. The very surface layer of the epidermis consists of the stratum corneum, a layer consisting of dead cells that are continually lost by attrition, and replaced in a dynamic process as the cells arise in the dermis and grow outwards.

Wound dressings and artificial skin are used to repair skin damage caused by bed sores, burns, cancer excision and complications of conditions such as diabetes (diabetic skin ulcers) and insufficient output from the heart (venous statis skin ulcers). Of these, bed sores are becoming more frequent, due to the increased numbers of patients living well into old age, but becoming progressively bedridden or seriously ill for extended periods of time. Bed sores, properly known as decubitis ulcers, have a number of causes, including reduction in capillary blood flow to the skin, infection and lack of sensation in the skin.[78] Skin ulcers of this type commonly occur at various points in the lower half of the body, including the hips, buttocks, knees, ankles and heels. Bacterial contamination may occur, complicating the treatment of the original condition. These ulcers can be treated by a number of approaches, including cold compresses, barrier dressings and disinfectant washes.

Ulcers may occur with diabetes or cardiac dysfunction. These arise as a consequence of circulatory problems. In diabetes, a result of the presence of excess glucose in the blood is the formation of a glycated derivative of collagen in the blood vessel walls. This leads to thickening of these walls, with resulting premature arteriosclerosis. This, in turn, leads to the development of skin ulcers that usually prove difficult to heal. A variety of dressings have been used for treating these ulcers, many of which are of biological origin, typically collagen sponges and other reconstituted collagen products.

In the case of severe burns, a variety of wound treatments have been developed that are based on synthetic polymer materials. Re-establishment of a sound natural skin is enhanced by keeping wounds wet, so that dressings are designed with the aim of maintaining moisture within the wound.[79] In addition, they should be flexible to cover the wound surface completely and be antiseptic and haemostatic.[80]

Polymers that have been used in wound dressings for burns are listed in Table 1.5.

Tissue engineering has been employed in a number of experimental studies of possible wound dressings and skin substitutes for burns and other wounds. For example, collagen sponge has been cultured with autologous fibroblasts and basal cells.[81] This developed an appropriate skin analogue within three weeks of placement. Rat-tail collagen seeded with autologous fibroblasts has also been used to produce a substitute skin for treatment of extensive injuries

Table 1.5 *Synthetic polymers used in wound dressings*

Polyurethane (used with adhesive)
Composite films of polyurethane with carboxymethylcellulose or polyethylene
Poly(2-hydroxyethylmethacrylate)
Polyurethane grafted with acrylamide
Silicone membrane on nylon velour
Poly(ethylene oxide) bonded to polyethylene mesh

---

in studies using rats as experimental models.[82] However, this did not result in the formation of full skin, since after 13 months, the resulting epidermis was still smooth, hairless and lacking sebaceous glands. It was also thinner than normal skin.

Overall, despite considerable research, attempts to prepare a widely useable artificial skin have not yet succeeded.[14] Skin grafting is widely used, despite the fragility of the grafts and other problems with the technique, and it seems likely that future developments in this area will employ tissue engineering techniques to prepare semi-synthetic treatments for burns and ulcers.

Facial Implants

Facial deformity may arise either from congenital disorders, severe trauma or the excision of tumours, and may involve the jaws, skull or face.[83] Congenital disorders include cleft lip and palate, and microtic ear. The latter is the malformation of the external part of the ear, so that only a small flap or significantly mis-shapen structure is present. It usually occurs asymmetrically, *i.e.* only one ear is affected.

Cleft lip and palate are dealt with surgically, with the clefts being stitched or repaired, sometimes with autografts. Other deformities make use of synthetic materials. For example, microtic ears can be replaced by silicone prostheses mounted on partially implanted abutments. The ear is typically modelled from the normal ear of the patient, and is mounted on a metal support, typically fabricated from a titanium alloy. These implants require careful management, since they penetrate the skin. They should, for example, be kept scrupulously clean by the patient, preferably with the aid of a bactericidal soap. Antifungal and antibacterial creams may also need to be applied.

In the case of severe trauma, it may be necessary to place an implant-stabilised facial prosthesis, possibly extending to contain a false eye. The basic principles of preparing such prostheses are similar to those for replacement ear. The titanium alloy needs sufficient bone density for sound anchorage, and the implant must be maintained with a good standard of hygiene. The major part of the prosthesis is mounted on a metal piece or pieces which clip onto the implant for retention.

Facial prostheses of this kind require careful monitoring. It is important that the colour should be stable, but exposure to sunlight or to smoky atmospheres may cause yellowing, so that the colour match with the natural

Table 1.6 *Types of breast implant*

Smooth gel-filled
Textured gel-filled
Polyurethane covered
Saline-filled

skin should be reassessed from time to time. Similarly, the tissues adjacent to the prosthesis must be examined regularly, since they often have reduced sensation and may become affected by the prosthesis without the patient noticing. Despite these problems, psychologically these prostheses are enormously beneficial for the patient, whose lives are often dramatically transformed by having them fitted.

Breast Implants

Prosthetic breasts are mainly used to increase size, the principal reason being cosmetic.[84] A significant number, however, are implanted for reconstruction following removal of breast tissue for health-related reasons, typically cancer. Breast implants come in a variety of designs, as listed in Table 1.6, but all share the feature of comprising a liquid inside an impervious sac. Of these, the smooth gel-filled implant is the most widely used.

The gel employed has typically been silicone fluid, but sufficient concerns have been expressed about leakage that these materials are no longer approved by the United States Food and Drugs Administration. Saline-filled implants have similar problems of leakage, even in the absence of obvious defects in the polymeric outer sac. However, saline does not cause any adverse reaction in the surrounding tissue, unlike silicone fluid, so small amounts of such leakage can be tolerated.[85] On the other hand, extensive leakage causes the implant to deflate, and then it must be replaced.

There are a number of complications that may arise with breast implants. These mainly relate to the development of fibrous capsule around the implant. In certain cases, this may become calcified, requiring further surgical intervention or removal of the implant.[86] The fibrous capsule may contract, resulting in disfigurement, implant migration and/or gross leakage.[87] Such contraction is significantly more likely with silicone-filled implants than with saline-filled ones, due to leakage of the silicone fluid in the former case.[88] Finally, the presence of fibrous capsule makes detection of breast cancer difficult,[89,90] and though there is no long term indication of implants increasing the incidence of breast cancer[91] women with implants who develop breast cancer generally have more advanced disease than those without implants.

Ear Ossicles

The ear consists of three sections, the outer, the middle and the inner ear. Sound is detected as the result of small waves of pressure in the air entering the

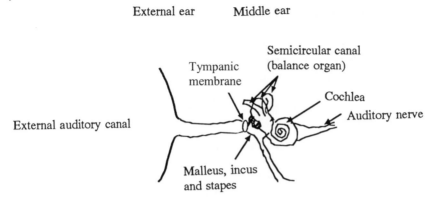

Figure 1.6 *Structure of the ear*

outer ear and causing slight movements of the tympanic membrane (see Figure 1.6). These movements are transmitted along the ossicular chain, which comprises three tiny, lightly anchored bones, called respectively the malleus, the incus and the stapes. Following transmission, the movement arrives at the oval window, a membrane that separates the middle ear from the inner ear, and here the movement becomes converted into electrical signals in the auditory nerve. These pass to the brain, where they are registered as sound.

Disease may lead to loss or damage to the ear ossicles and one solution is to employ biomaterials as ossicular prostheses. Implants have also been used to replace the bony auditory canal wall which may have to be removed in radical mastoid surgery, though these have not been as widely used as artificial ossicles.

In general it is possible to achieve good early results in restoration of hearing with synthetic materials. On the other hand, longer term success is less easy, because implanted ossicles may suffer from infection, displacement and extrusion. A variety of materials have been used, including polyethylene, poly(tetrafluoroethylene) and ceramics such as sintered hydroxyapatite[92] and glass-ceramics.[93] More recently, composites of polyethylene filled with hydroyapatite have been employed. These have mechanical properties that closely mimic those of natural mineralised tissue, and this seems to have reduced the problems of displacement and extrusion.[94] Infection is a function of the skill of the individual surgeon and, as so often in this field, the more skilled the otological surgeon, the more likely are satisfactory and long-lasting outcomes.[95]

3 Conclusions

This chapter has outlined some of the most common clinical uses of synthetic materials. The purely chemical aspects of these materials are rarely considered as separate topics, because the field is so strongly interdisciplinary, involving as it does the expertise of engineers, biologists and surgeons, as well as chemists. On the other hand, many of the problems in the field are essentially chemical, whether arising from the composition of the material, or from surface chemical effects. Success with implantable materials cannot be achieved without an understanding of the appropriate chemistry, and in the chapters that follow, both the chemistry of the interaction of materials with the body, and of their composition and fabrication are described.

4 References

1   J. Charnley, *J. Bone Jt. Surg.*, 1960, **42B**, 28.
2   J. Charnley, *Lancet*, 1961, 1129.
3   D. F. Williams in *Progress in Biomedical Engineering*, Vol. 4, ed. D. F. Willaims, Elsevier Science Publishers, London, 1987.
4   J. Charnley, *J. Bone Jt. Surg.*, 1964, **46B**, 518.
5   J. Charnley, *J. Bone Jt. Surg.*, 1970, **50B**, 340.
6   D. F. Williams, in *Fundamental Aspects of Biocompatibility*, Vol. 1, CRC Press, Boca Raton, FL, 1981, Chapter 2.
7   R. van Noort, *J. Mater. Sci.*, 1987, **22**, 3801.
8   G. Meacim and D. F. Williams, *J. Biomed. Mater. Res*, 1973, **7**, 555.
9   J. T. Scales, *J. Bone Jt. Surg.*, 1991, **73B**, 534.
10 S. B. Goodman, V. L. Fornasier, J. Lee and J. Kei, *J. Biomed. Mater. Res.*, 1990, **24**, 1539.
11 B. Kasemo and J. Lausmaa, *CRC Crit. Rev. Biocompatibility* 1986, **4**, 335.
12 L. L. Hench, *Biomaterials*, 1998, **19**, 1419.
13 L. A. Flemming, *J. Gen. Intern. Med.*, 1992, **7**, 554.
14 R. B. Mazess, *Clin. Orthop. Rel. Res.*, 1982, **165**, 239.
15 L. Mosekilde, *Bone*, 1995, **14**, 3435.
16 L. J. Melton, *J. Bone Min. Res.*, 1995, **10**, 175.
17 R. Sagraves, *J. Clin. Pharmacol.*, 1995, **35**, 2S.
18 J. R. Phelps and R. A. Dormer, *Handbook of Biomaterials Evaluation*, ed. A. F. von Reccum, Macmillan Publishing Corp., New York, 1986, p. 503.
19 F. H. Silver, *Biomaterials, Medical Devices and Tissue Engineering: An integrated approach*, Chapman and Hall, London, 1994.
20 J. L. Lewis and W. D. Lew, *Handbook of Bioengineering*, ed. R. Skaka and S. Chien, McGraw-Hill, New York, 1987, Chapter 40.
21 M. Long and H. J. Rack, *Biomaterials*, 1998, **19**, 1621.
22 V. C. Mow and L. J. Soslowsky, *Basic Orthopaedic Biomechanics*, Raven Press, New York, 1991, p. 245.
23 J. Charnley, *Clin. Orthop.*, 1972, **87**, 167.
24 J. Delécrin, S. Szmuckler-Moncler, G. Daculsi, J. Rieu and B. Duquet, in *Bioceramics*, ed. W. Bonfield, G. Hastings and K. E. Tanner, Butterworth-Heinemann, Oxford, UK, 1991, Vol. 4, p. 311.
25 I. Oh and W. H. Harris, *J. Bone Jt. Surg.*, 1978, **60A**, 75.
26 J. Jowsey, *Clin. Orthop.*, 1960, **17**, 210.
27 C. A. L. Bassett, *Cal. Tissue Res.*, 1968, **1**, 252.
28 J. H. Dumbleton, in *Joint Replacement Arthroplasty*, ed. B. F. Morrey, Churchill Livingstone, New York, 1991, Chapter 6.
29 H. McKellop, I. C. Clarke, K. L. Markolf and H. C. Amstutz, *J. Biomed. Mater. Res.*, 1981, **15**, 619.
30 R. M. Hall and A. Unsworth, *Biomaterials*, 1997, **18**, 1017.

31 F. Silver and C. Doillon, *Biocompatibility, Vol. 1: Polymers*, VCH Publishers Inc, New York, 1989.

32 J. Black, *Orthop. Clin. N. Am.*, 1989, **20**, 1.

33 J. D. Currey, *Proc. Inst. Mech. Eng.*, 1998, **212**, Part H, 399.

34 C. C. Danielson, L. Mosekilde and B. Svenstrap, *Calcif. Tissue Int.*, 1993, **52**, 26.

35 R. W. McCalden, J. A. McGeogh, M. B. Baker and C. M. Court-Brown, *J. Bone Jt. Surg.*, 1993, **75A**, 1193.

36 A. H. Burnstein, D. T. Reilly and M. Martens, *J. Bone Jt. Surg.*, 1976, **58A**, 82.

37 K. Ishikawa, S. Takagi, L. C. Chow and Y. Ishikawa, *J. Mater. Sci. Mater. Med.*, 1995, **6**, 528.

38 S. Lees and C. L. Davidson, *J. Biomechanics*, 1977, **10**, 477.

39 S. A. Reid, *Anat. Embryol.*, 1986, **174**, 329.

40 V. Ziv, H. D. Wagner and S. Weiner, *Bone*, 1996, **18**, 1.

41 M. J. Friedman and R. J. Ferkel, *Prosthetic Ligament Reconstruction of the Knee*, W. B. Saunders Co., Philadelphia, PA, 1988.

42 B. Zarims and M. Adams, *New Eng. J. Med.*, 1988, **318**, 950

43 C. Frank, S. L.-Y. Woo, D. Amiel, F. Harwood, M. Gomez and W. A. J. Atkinson, *Am. J. Sports Med*, 1983, **11**, 379.

44 H. Alexander, J. R. Parsons, I. D. Strauchler, S. F. Corcoran, O. Gona and A. B. Weiss, *Orthop. Rev.*, 1981, **X**, 41.

45 A. E. Goodship, S. A. Wilcock and J. S. Shah, *Clin. Orthop.*, 1985, **196**, 61.

46 E. H. Chen and J. Black, *J. Biomed. Mater. Res*, 1990, **14**, 567.

47 L. M. Buja, The vascular system, in *Basic Pathology*, ed. S. L. Robbins, V. Kumar and W. B. Saunders, Philadelphia, PA, 1987, Chapter 10.

48 *The United Kingdom Heart Valve Registry*, Department of Health, London, 1995.

49 D. Larenby and J. P. Gold, *Infect. Med.*, 1991, **8**, 15.

50 D. K. San, S. Y. Jeong, Y. H. Kim and B. G. Min, *J. Biomater. Sci. Polymer Edn.*, 1992, **3**, 229.

51 D. F. Williams, *Blood Compatibility*, Vol. 1, ed. D. F. Williams, CRC Press, Boca Raton, FL, 1982.

52 R. Macnair, M. J. Underwood and G. D. Angelini, *Proc. Inst. Mech. Eng.*, 1998, **212**, *Part H*, 465.

53 B. Kasemo and J. Lausma, *Environ. Health Perspect.*, 1994, **102**, 41.

54 F. H. Silver and D. A. Swann, *Int. J. Biol. Macromol.*, 1982, **4**, 425.

55 R. L. Cleland and J. L. Wang, *Biopolymers*, 1970, **9**, 799.

56 A. Tardieu and M. Delaye, *Ann. Rev. Biophys. Chem.*, 1988, **17**, 47.

57 D. P. DeVore, *J. Long-term Effects Med. Implants*, 1991, **1**, 205.

58 D. J. Apple, R. N. Brems, R. B. Park, D. Kavka-Van Norman, S. O. Hansen, M. R. Tetz, S. C. Richards and S. D. Letchinger, *J. Cataract Refract. Surg.*, 1987, **13**, 157.

59 D. J. Apple, N. Mamalis, K. Lotfield, J. M. Googe, L. C. Noval, D. Kavka-Van Norman, S. E. Brady and R. J. Olson, *Survey Ophthalmol.*, 1984, **29**, 1.

60 R. L. Peiffer Jr, *J. Cataract Refract. Surg.*, 1987, **13**, 397.

61 N. L. Burnstein, M. Ding and M. V. Pratt, *J. Cataract Refract. Surg.*, 1988, **14**, 520.

62 J. I. Lippman, *Contact Lens Assoc. Ophthalmologists' J.*, 1990, **16**, 287.

63 W. A. Franks, G. G. W. Adams, J. K. G. Dart and D. Minassian, *Br. Med. J.*, 1988, **297**, 534.

64 A. C. B. Molteno, *Br. J. Ophthal.*, 1969, **53**, 161.

65 A. W. Lloyd, R. G. A. Faragher and S. P. Denyer, *Biomaterials*, 2001, **22**, 769.

66 F. J. Holly and M. A. Lemp, *Surv. Opthalmol.*, 1977, **21**, 69.

67 M. B. Limberg, C. McCaa, G. E. Kissling and H. E. Kaufman, *Am. J. Ophth.*, 1987, **103**, 194.

68 T. R. Pitt Ford, *The Restoration of Teeth*, Blackwell Scientific, Oxford, UK, 1985.

69 S. Hojo, N. Takahashi and T. Yamada, *J. Dent. Res.*, 1991, **70**, 182.

70 R. M. Stephan, *J. Am. Dent. Assoc.*, 1940, **27**, 718.

71 J. W. Nicholson, *Int. J. Adhesion Adhesives*, 1998, **18**, 229.

72 D. M. Davis, *Int. J. Prosthetics*, 1990, **3**, 42.

73 W. R. Laney and D. E. Tolman, in *The Brånemark Osseointegrated Implant*, ed. T. Albrektson and G. A. Zarb, Quintessence Books, Chicago, IL, 1989, pp. 165–195.

74 R. van Noort, *J. Mater. Sci.*, 1987, **22**, 3801.

75 P. N. Calgut, I. M. Waite and S. M. B. Tinkler, *Clin. Mater.*, 1990, **6**, 105.

76 J. C. Geeson and R. A. Berg, Biochemistry of skin, bone and cartilage, in *Applications of Biomaterials in Facial Plastic Surgery*, ed. A. I. Glasgold and F. H. Silver, CRC Press, Boca Raton, FL, 1991, Chapter 2.

77 A. J. Wasserman and M. G. Dunn, Morphology and mechanics of skin, cartilage and bone, in *Applications of Biomaterials in Facial Plastic Surgery*, ed. A. I. Glasgold and F. H. Silver, CRC Press, Boca Raton, FL, 1991.

78 L. C. Parish, J. A. Witkowski and J. T. Crissey, *The Decubitis Ulcer*, Masson Publishing USA Inc., New York, 1983.

79 D. T. Rovee, *Clin. Mater.*, 1991, **8**, 83.

80 B. G. MacMillan, in *Burn Wound Coverings*, Vol. 1, ed. D. L. Wise, CRC Press, Boca Raton, FL, 1984, p. 115.

81 C. J. Doillon and F. H. Silver, *Biomaterials*, 1986, **7**, 3.

82 E. Bell, *Plast. Reconstr. Surg.*, 1981, **67**, 386.

83 J. A. Hobkirk and R. M. Watson, *Color Atlas of Dental and Maxillo-Facial Implantology*, Mosby-Wolfe, London, 1995.

84 J. Sprague Zones, *J. Long-term Effects Med. Implants*, 1992, **1**, 225.

85 C. Batich and D. De Palma, *J. Long-term Effects Med. Implants*, 1992, **1**, 253.

86 A. B. Redfern, J. J. Ryan and T. C. Su, *Plast. Reconstr. Surg.*, 1977, **59**, 249.

87 B. Brandt, V. Breiting, B. L. Christensen, M. Neilsen and J. L. Thomson, *Scand. J. Plast. Reconstr. Surg.*, 1984, **18**, 311.

88 O. Aspland, *Plast. Reconstr. Surg.*, 1984, **73**, 270.

89 H. Hayes Jr, J. Vandegrift and W. C. Drier, *Plast. Reconstr. Surg.*, 1988, **82**, 1.

90 D. D. Dershaw and T. A. Chaglassian, *Radiology*, 1989, **170**, 69.

91 M. J. Silverstein, N. Handel and P. Gamgami, *Arch. Surg.*, 1985, **123**, 681.

92 J. J. Grote, *Arch. Otolaryngol.*, 1984, **110**, 197.

93 R. Reck and J. Helms, *Am. J. Otol.*, 1985, **6**, 280.

94 K. E. Tanner, R. N. Downes and W. Bonfield, *Br. Ceram. Trans.*, 1994, **93**, 104.

95 R. L. Blain, in *Recent Advances in Otolaryngology*, No. 6, ed. R. F. Gray and J. A. Rutka, Churchill-Livingstone, Edinburgh, 1988, Chapter 13.

CHAPTER 2
Polymers

1 Polyethylene

Polyethylene is a predominantly linear polymer of the monomer ethene (ethylene), *i.e.*:

$$-(CH_2CH_2)_n-$$

It is currently the largest tonnage plastic material in the world, having first been produced commercially as long ago as 1939, when it was first marketed by ICI in the UK as an electrical insulator.[1] The most common grades have number average molecular weights in the range 10 000–40 000, though there are a number of special purpose grades on either side of this range.[2] In particular, the special grade known as ultrahigh molecular weight polyethylene, UHMWPE, used in medical implants, has molecular weights in the range of millions. These grades are prepared by the Ziegler process, *i.e.* using co-ordination reactions with metal alkyls as catalysts.[1] They are difficult to process because they cannot be melted without decomposition, and must be formed into useful articles in the rubbery phase.[2]

Ultrahigh molecular weight polyethylene, UHMWPE, is defined by ISO 11542 as having a molecular weight greater than 1×10^6; by contrast, the ASTM D 4020 definition states that it is a linear polymer with a molecular weight greater than 3.1×10^6. However, the nomenclature of this group of polymers has changed considerably in the last 30 years, as explained by Kurtz *et al.*,[3] making comparisons of clinical results across the years difficult. Whatever the precise details of the molecular weight, the material consists of predominantly linear molecules that, because of their large size, are relatively poorly crystalline. Consequently, they are of very low density, typically of the order of 0.94 g cm^{-3}, and properties that are dependent on degree of crystallinity, such as yield strength and modulus, tend to be slightly worse than those of polyethylenes of more conventional molecular weight.[2] On the other hand, they have very high abrasion resistance, though wear is a problem, as discussed later in the chapter. UHMWPE also possesses good impact strength, shows reasonable resistance to stress-cracking and relatively low creep, all of which are advantageous when this material is used in medical devices.

The history of the use of UHMWPE in orthopaedics began in 1962 when Charnley began to employ this material in total hip arthroplasties.[6] At the time, he referred to the material simply as high density polyethylene, HDPE, a term that was correct at the time[3] but which is no longer considered correct. Currently, HDPE is defined as having a molecular weight below 2×10^5, *i.e.* an order of magnitude less than UHMWPE under the modern definitions. The brand of material actually used by Charnley was RCH 1000, manufactured by Hoechst, produced initially as compression moulded sheets, and used to fabricate the acetabular cap components.

Since the introduction of this material to orthopaedics, polymers have undergone several changes in brand name, and other manufacturers, such as Hercules Powder and Montell, have also produced grades of UHMWPE suitable for clinical use. Molecular weights lie in the range 3.5 to 6×10^6, and a number of the technical grades available incorporate calcium stearate to act as a radical scavenger and lubricant.

Determination of molecular weight is important for these materials, because of its influence on the mechanical behaviour. Routinely, intrinsic viscosity measurements have been used, though there are difficulties in dissolving the polymer, even at elevated temperatures, and these pose severe practical limitations on the characterisation of UHMWPE by solution methods[3]. Size-exclusion chromatography has also been used,[4] employing trichlorobenzene as solvent at 145 °C. Under these conditions, an antioxidant, such as *N*-phenyl-2-naphthylamine is needed.[3]

Molecular weight is not the only important parameter, since crystallisation can also affect physical properties. The precise effect is poorly understood, partly because there has been relatively little work on this topic, but it is known that materials from different manufacturers have different crystal morphologies. For example, the technical grade UHMWPE Montell 1900 has been shown to exhibit spherulitic crystalline morphology,[5] whereas polymers from Hoechst tend to have a lamellar morphology.[6] These differences may explain differences in behaviour in a laboratory wear test. Using bidirectional pin-on-disk testing, it has been found that the Hoechst UHMWPE lost 7.87–9.88 mg per million cycles, while the Montell polymer lost between 8.68 and 9.36 mg per million cycles.[7] Thus processing, with the effect this has on the crystallisation process, may play an important role in determining the wear behaviour of UHMWPE.

Calcium Stearate in UHMWPE

Calcium stearate, $[CH_3(CH_2)_{16}COO^-]_2Ca^{2+}$, has been used as an additive for technical grades of polyethylene since 1955, when the Hoechst company introduced it.[3] It has several functions: it acts as a scavenger for catalyst residues that might otherwise remain in the polymer and promote corrosion in the processing equipment. It also has lubricating properties and acts as a release agent. Calcium stearate is a waxy solid that surface-coats the polymer particles when employed in polyethylene production.

The presence of calcium stearate seems to cause subtle changes in the performance of UHMWPE in total hip replacement.[8,9] For example, it is associated with fusion defects, accumulating at grain boundaries and altering a range of properties, including ultimate tensile strength, elongation to failure, fracture toughness and fatigue resistance.[10,11] The mechanism whereby these properties are altered seems to be that the geometries of the fusion defects and other internal microvoids are modified.[3,9,12] Since these imperfections have been shown by a fracture mechanics approach to be able to act as stress concentrators and to promote the nucleation of cracks,[13] it seems likely that their alteration and enlargement would lead to the observed reduction in the various physical properties.

Although calcium stearate seems to increase the problems of fusion defects, it is not clear how this contributes to wear in acetabular components. There is some evidence that reducing the level of calcium stearate leads to a reduction in wear. Schmidt and Hamilton[11] showed a reduction in wear from 15.9 ± 2.1 mg/10^6 cycles to 10.9 ± 1.9 mg/10^6 cycles for two grades of UHMWPE containing calcium stearate and without calcium stearate respectively, using a hip simulator. In another study, McKellop *et al.*[14] came to different conclusions. However, they employed grades of polyethylene that differed in molecular weight as well as calcium stearate content. In their hip simulation experiments they found that the decreased molecular weight and reduced calcium stearate material did not have improved wear characteristics. As Kurtz *et al.*[3] pointed out, though, since the calcium stearate content was not varied independently of molecular weight, a definite conclusion regarding the effects of calcium stearate alone on wear was precluded.

The presence of calcium stearate has been shown to enhance oxidation of polyethylene in accelerated ageing tests.[10] It also seems to influence the oxidation resistance of the polymer following gamma ray sterilisation, though the mechanism for this is not clear. However, in the light of these findings, a number of suppliers of orthopaedic devices are switching to grades of UHMWPE that contain reduced levels of calcium stearate.

Polyethylene for Orthopaedics

As previously explained, UHMWPE is widely used in artificial joints as a load-bearing material.[15,16] Originally introduced in 1962[17] as the acetabular cap component of the low friction hip joint, more recently this material has been used as the tibial plateau component of the artificial knee joint. These devices thus comprise metal-on-UHMWPE bearings. Other materials have been used, such as metal-on-metal and ceramic-on-ceramic, but joints involving UHMWPE remain the most widely used in modern surgery.[3] In the short term UHMWPE has excellent physical and chemical properties, and is acceptable to the body, showing good biocompatibility when cemented into the bone with a self-curing poly(methyl methacrylate) bone cement.

In the longer term, however, there are problems with this material. For example, it will creep and shows fatigue fracture. It is also subject to wear, a problem that is increasingly being recognised as a major one with all artificial joints that rely on the articulation of UHMWPE with metal or ceramic components. In a total hip replacement, the wear rate of the UHMWPE component in patients is typically less than 0.1 mm per year. At this rate, it would take 100 years to wear away a typical 10 mm thick UHMWPE component.[18] Hence it is not loss of the component which poses the problem, but the effects of the debris around the joint. This particulate material has been found to lead to osteolysis and implant loosening under certain circumstances.[3] Creep, too, is an important problem, and is significant because, although it occurs to a much lesser extent in UHMWPE than in low molecular weight polyethylenes, it nonetheless occurs to a much greater extent than in the metal or ceramic components of the artificial joint, or than in cortical bone. Creep can alter patterns of stress transfer, which in turn can alter the body's ability to maintain appropriate bone density adjacent to the implant. An urgent goal of current research in this field is the reduction in both wear and creep, thereby improving the long-term perfomance of replacement hip devices, particularly for heavier, younger and more active patients.[19]

The chemical stability of polyethylene in the body is good, but there are changes in properties with time, and these are consistent with chemical degradation taking place. Inside the body, polyethylene is protected from the major agents which bring about degradation, such as elevated temperatures, oxygen, gamma radiation or UV light.[20] However, it has often been stabilised by gamma irradiation prior to placement, and this introduces oxygen to the material, generally in the form of carbonyl groups. These, in turn, render the material susceptible to enzymic attack, and it has been shown that, over time, there is a decrease in molecular weight, an increase in density, an increase in crystallinity and an increase in carbonyl oxidation products in UHMWPE in orthopaedic applications.[21]

The key stage in making polyethylene susceptible to degradation is photo-oxidation. This is a free-radical process and is promoted by gamma sterilisation. It causes a reduction in molecular weight and the introduction of oxygen into the structure,[22] an effect which itself accelerates the degradation processes.[23] Within the body, polymer degradation tends to be mediated typically by enzymes that promote hydrolysis, and this is enhanced by the presence of polar functional groups, typically those containing heteroatoms such as oxygen. Even without such sterilisation polyethylene is capable of undergoing degradation inside the body. For example, some years ago it was shown in an investigation of the carcinogenic properties of polymers in animal models that polyethylene underwent degradation.[24] The investigation used a number of ^{14}C labelled polymers, including polyethylene, and in all cases it was possible to detect labelled products in the urine, faeces or respiratory CO_2. For polyethylene, a time of 26 weeks was needed for these degradation products to appear.

The possibility of this type of degradation has become of significant concern

Table 2.1 *Raman spectroscopy of UHMWPE with and without gamma irradiation (after Chenery[29])*

Form	Gamma ray dose/kGy	Bands of oxidised species/cm^{-1}	Assignment
Sheet	Not	None	
Powder	Not	None	
Sheet	25	870–873	Peroxy
		933–935	Epoxy
		1151	Alcohol
		1770	Peroxy acid
Powder	25	None detected (but carbonyl in IR)	

in recent years, and there are now worries that, when polyethylene is used as a biomaterial, it is not as homogeneous and stable throughout the lifetime of the component as was once assumed. There has now been a change in sterilisation methods employed in implant manufacture and a move away from gamma irradiation. However, as we have seen, the fact that most polyethylene devices for use in orthopaedics have been sterilised by gamma irradiation prior to placement means that there are many patients around the world with implants containing a significant oxygen content in the surface layers, at least.[25] The concern now is that these oxygenated functional groups will increase the susceptibility of these implants to enzymic degradation and also alter the mechanical properties, possibly accelerating wear and the production of particulate debris, with all the biological problems this causes.[26]

The mechanism of reaction subsequent to gamma irradiation appears to be as follows.[27] A gamma photon is absorbed, causing the formation of an excited state. This decays *via* homolytic fission of either C–C of C–H bonds to yield free radicals. Because these radicals are effectively matrix isolated within the polyethylene, they are quite long-lived. Oxygen from the atmosphere can therefore diffuse in and, once at the free radical site, can undergo reaction, leading to the formation of carbonyl species and further chain scission.

Appropriate carbonyl groups have been shown to be present in both new and retrieved UHMWPE using FTIR spectroscopy.[26] In addition, free radicals have been detected using electron spin resonance.[28] Raman spectroscopy has also been employed to study the effects of gamma irradiation,[29] this technique being prefered because many of the likely functional groups are not formally active in the infrared, but do give an absorption in the Raman. Peroxy groups (C–O–O–C), for example, have been detected using the latter technique, yet if the atoms are completely coplanar this group gives no absorption in the infrared. By contrast, the very symmetry of the O–O bond makes it a good Raman chromophore. With the aid of Raman spectroscopy, a variety of oxygenated functional groups have been detected, as shown in Table 2.1.

It was found that irradiation at 50 kGy did not lead to any systematic differences in the spectra. The fact that peroxy species were detected in the powder was attributed to them decomposing to ketones more rapidly than in the sheet, presumably because oxygen could diffuse more readily into the surface layers of the high surface area powder than it could into the sheet.

The overall conclusion to be drawn from these studies is that it is the gamma sterilisation process that is responsible for the development of oxygenated functional groups within UHMWPE. This follows from the observation that there was no evidence of oxidation in the unsterilised UHMWPE blanks.

That these chemical changes do not influence only the enzymic degradation of the UHMWPE but also the mechanical properties has been demonstrated experimentally. For example, in a study of tibial plateau components of knee prostheses, Pascaud et al.[30] showed that sterilisation with gamma rays led to a 50% reduction in fracture toughness and a decrease in tearing modulus of 30%. Such a severe reduction in mechanical properties could in turn lead to a major reduction in the residual strength of components, and a significant shortening of the service life. These differences in fracture behaviour were not matched by comparable changes in static mechanical properties, since both Young's modulus and ultimate tensile strength were reduced only by about 7% in each case. Thus, the ability of cracks to propagate through the material seems to be the feature mainly affected by the occurrence of surface oxidation reactions.

Various studies have shown that the density of UHMWPE is not stable, but changes both on gamma irradiation and exposure to the body. In a very comprehensive study of UHMWPE in orthopaedics, Bostrom et al.[21] examined the effect of irradiation and long-term exposure to the body of tibial inserts using two different grades of polymer, some samples of which were stored for a considerable time on the shelf; other specimens were retrieved from patients. Retrieval was possible because the implants became associated with a variety of complications in patients, including infection, loosening, instability and chronic synovitis. From each of the specimens available, cores of material were obtained, and the density measured using a density column of isopropanol and water. Fourier Transform Infrared (FTIR) spectra were also recorded from all specimens.

The density of all sterilised and retrieved samples was found to be greater than that of virgin material, with outer layers showing greatest increase in density. Age of the implant was also found to influence density. These increases, which arise not only from the effects of gamma irradiation and subsequent chain scission, but also from fluid absorption,[31] are important because correlations have been found from *in vivo* experiments that a small increase in density, from 0.9350 to 0.9450 g cm^{-3} are accompanied by an increase in modulus of at least 25%.[32] A change of this magnitude near the articulating surface of a total joint component would cause significant increases in the stresses associated with wear damage at or near these surfaces.

The density increases due to a complex sequence of events that follow sterilisation. These continue for considerable periods of time after the initial sterilisation event and occur as a result of chemical reactions involving oxidation processes during irradiation. Because of the importance of oxygen to these reactions, diffusion has an important effect on the extent and rate of reaction and density changes tend to be greatest closest to the surface, where oxygen is more readily available.

FTIR confirmed that oxidation had occurred, with clear evidence of keto group formation from the carbonyl band at approximately 1720 cm^{-1}, and also of ester group formation from the band at about 1740 cm^{-1}. The precise mechanism of formation of these carbonyl groups is not clear but their occurrence has been found to vary with specimen history. Keto carbonyl groups have been found in gamma irradiated UHMWPE,[33] whereas the ester carbonyl group has been observed only in retrieved implants,[34] and has not been found in specimens aged outside the body.[31] Thus ester group formation seems to be dependent on enzymic modification of the polymer following interaction with body fluids, and is not the result of a simple non-biological oxidation process.

Overall, these findings show that gamma irradiation leads to significant chemical changes in UHMWPE and that these lead to continued chemical reaction following implantation into patients. These changes, in turn, adversely affect the physical properties of the device, and hence have an influence on the useful lifetime of the implant. However, until 1995 gamma ray sterilisation was considered the method of choice, which means that there are large numbers of such sterilised components currently in patients, a situation that will require careful scrutiny in the years ahead.

Processing of UHMWPE for Use in Orthopaedics

UHMWPE components are generally fabricated by manufacturing companies that are specialists at polymer processing, but are not the primary producers of the polymer. These companies buy in stock material of UHMWPE, either as beads or as preformed shapes, such as slabs or rods, and they use these forms of UHMWPE as the raw material from which they produce the finished orthopaedic components.

The shape of the final component may influence the choice of UHMWPE stock. For example, preformed slabs are relatively easy to mill into the reasonably flat tibial components of the artificial knee joint. Also, this is advantageous for the long-term survival of the knee joint, because preformed polymer slab tends to have a uniform polymer morphology[35] and consequently is more isotropic in its resistance to crack propagation than polymer that has been moulded from beads.[36] It therefore survives better under clinical service conditions.

Processing of preformed UHMWPE slabs involves various milling and turning operations. Details of processing techniques vary from supplier to supplier and are not usually disclosed in public because they are proprietary. However, it is known that they often involve the use of pre-shaped stock that is finished under controlled conditions of feed rate, cutting force and spindle speed. Close control of these steps is necessary because UHMWPE can be damaged by excessive heating. Current generation milling machines are capable of speeds of up to 12 000 rpm, and overheating of the polymer at this

Table 2.2 *Comparison of the properties of different grades of UHMWPE (after Kurtz et al.[16])*

Polymer grade	Hoechst GUR 1120	Hoechst GUR 1150
Molecular weight	3.5×10^6	$5.5 - 6 \times 10^6$
Processing	Compression moulded	Ram extruded
Impact strength (KJ m^{-2})	179	134
Yield stress (MPa)	23.3	21.7
Failure stress (MPa)	238	187
Fracture toughness (MPa m$^{\frac{1}{2}}$)	4.7	3.6

speed is avoided by proper control of feed rate and cutting force during machining.

The alternative to machining preformed UHMWPE stock is for the manufacturer to buy in UHMWPE beads or pellets and subject these to a consolidation process. For some time, primary manufacturers supplied only lower molecular weight UHMWPE for compression moulding by component manufacturers, and prepared higher molecular weight UHMWPE in preformed shapes, for example as extruded rod. Literature studies that have considered the effect of processing conditions only on the properties and performance of UHMWPE components, while ignoring the fact that molecular weight also varies have missed an important variable. Nonetheless, there are significant differences in physical properties between the types of UHMWPE, though the origin of these differences is not defined. Data in Table 2.2 illustrate this point.

There is also a technique for converting UHMWPE that is effectively a hybrid of direct compression moulding and machining of preformed stock material, namely hot isostatic pressing. In this process, cylindrical compact rods of UHMWPE are produced by a cold pressing process that expels most of the air. Subsequently, these so-called 'green' rods are sintered in a hot isostatic pressure furnace under a low-oxygen atmosphere. This feature is necessary to prevent oxidative degradation of the UHMWPE. The resulting rod stock is isotropic, and similar in morphology and mechanical properties to compression moulded polymer. Final implant components are prepared from the processed rod by various turning or milling operations.

Direct compression moulding, sometimes known as net shape compressive moulding, remains important in the fabrication of UHMWPE components for orthopaedic surgery because it offers a number of advantages. One is that, because the component has not been subject to separate milling or turning operations, it has a smooth surface free from machining marks. Another is that the physical and mechanical properties of the component can be tailored to some extent by careful attention to the processing conditions during the moulding cycle.[37,38] On the other hand, process control of direct compression moulding is difficult, hence where possible implant manufacturers find it easier to work with preformed rod or sheet.

Sterilisation of UHMWPE

As was explained earlier, the link between gamma ray sterilisation and degradation of UHMWPE has been established[39] and has led to a change in practice since 1995, with the widespread adoption of alternative means of sterilisation.[3] As previously carried out, gamma ray sterilisation involved exposure of the component to a dose of between 25 and 40 kGy in air. By 1998, most manufacturers had changed completely and were using one of the following methods:

  (i) gamma irradiation in a reduced oxygen atmosphere;
  (ii) treatment with ethylene oxide gas;
  (iii) treatment in a gas plasma.

Of these, the first was the simplest to switch to, since it involves more or less unchanged sterilisation equipment. It is applied, though, to components that have been prepackaged under reduced oxygen conditions prior to irradiation. Gamma rays are effective in sterilisation because they not only eliminate bacteria, they also promote some cross-linking within the UHMWPE. This is an advantage, since cross-linking reduces wear of UHMWPE under service conditions.

A widely used alternative is sterilisation with ethylene oxide gas.[40] Ethylene oxide is highly toxic and destroys bacteria, spores and viruses.[3] It lends itself to sterilisation of UHMWPE because it does not react with the polymer, and there seem to be no physical, chemical or mechanical changes in the polymer during the sterilisation process.[41]

Sterilisation with ethylene oxide involves diffusion of the gas into the surface layers of the UHMWPE component over several hours[40] at slightly elevated temperatures, typically 46 °C. Following this, the residual ethylene oxide is allowed to diffuse out of the polymer to reduce the possibility of an adverse tissue reaction on implantation. The entire cycle typically takes 41 hours.

Retrieval studies suggest that ethylene oxide-sterilised UHMWPE has distinct advantages over material sterilised by gamma irradiation in air. In particular, retrieved specimens have been found to show less surface damage and delamination than gamma sterilised UHMWPE.[42]

The final option is gas plasma sterilisation. This is the most recent of the sterilisation techniques to become available and has not yet, at least, become particularly widely used. It involves exposing the UHMWPE component to a gas plasma, which may be either of peracetic acid or of hydrogen peroxide, typically at temperatures below 50 °C.[43] Current indications are that this technique, which is free from toxic residues or environmentally damaging by-products, causes no chemical or corresponding mechanical changes in the UHMWPE.[44] Since it is quicker and less hazardous than treatment with ethylene oxide gas, it has substantial advantages as a sterilisation technique. However, it is too soon for information to be available on the *in vivo* performance of gas plasma sterilised UHMWPE components, and additional research is necessary before it can be adopted as the method of choice for sterilising polyethylene components for total joint replacement.

Wear of UHMWPE

Wear of the bearing surfaces of UHMWPE is now recognised as one of the most important factors in determining the long-term clinical performance of orthopaedic joint replacements.[45] The reason for the concern is that wear causes the formation of particulate debris, and the accumulation of this debris leads to adverse biological responses, including inflammation and associated discomfort to the patient.

When retrieved joint components are examined it is often possible to observe gross changes, even with the naked eye. Typical detectable changes of this kind include discoloration, deformation, cracking, abrasion, delamination and perforation. The appearance of the component may vary depending on whether or not the surface was load-bearing. For example, in retrieved acetabular cups not previously exposed to gamma rays, weight-bearing surfaces have been found to be smooth and glossy, whereas in many cases the non-weight-bearing surfaces were degraded, being rough and yellow-brown in colour.[46]

There are also features of these retrieved components that need scanning electron microscopy (SEM) for their detection. SEM reveals a number of details not visible to the naked eye. For example, on non-weight-bearing portions of polyethylene components, it is possible to observe flaws and scratch marks introduced during the production process. Some parts showed irregular line patterns of ripples varying in size from one to several micrometres, features that were assumed to arise from degradation of the polymer within the body.[47]

Retrieval of polyethylene implants up to 14 years after placement has revealed changes in the weight-bearing areas, and these have been found to vary with the nature of the opposing material.[47] For example, polyethylene opposed by a 22 mm stainless steel femoral head showed an essentially smooth surface, which at higher magnification proved to be a carpet-like structure composed of fine fibres. Occasional scratches and fine crevices were also observed. By contrast, polyethylene opposed by a 32 mm cobalt–chromium–molybdenum alloy head showed a scale-like morphology at higher magnifications, with many scratches but few crevices. Finally for polyethylene opposed by a 28 mm 'Bioceram' (alumina) femoral head, there was a much rougher scale-like morphology visible at higher magnifications, and also fine crevices but no scratches. This study showed the importance of the total engineering system in determining the pattern of wear in the polyethylene component.

There have been studies of the particulate debris that occurs around artificial hip joints, and this has been found to comprise mainly submicron polyethylene.[48] The chemical nature of this debris has a significant effect on the way the body responds to it, and studies have shown that oxidised and virgin high density polyethylene provoke quite different reactions.[49] These include variations in the amount of the substances IL-1, TNF-α and IL-6 secreted in the region adjacent to the build-up of debris.[50] The precise impact of these differences on the details of the biological response is still being investigated.

The mechanisms of wear have been shown to be different in artificial hip and knee joints,[51] but in both cases wear seems to be related to the plastic flow behaviour of the UHMWPE. In acetabular components of the artificial hip, particles of wear debris are generally of submicron dimensions, and their formation seems to result from plastic strain under multiaxial loading conditions. This strain builds up until a critical strain is reached, whereupon the material begins to fragment.[52,53] In the artificial knee joint, the fatigue and fracture mechanisms have also been related directly to the plastic flow behaviour of UHMWPE as well as to such properties as the yield stress and ultimate stress.[54] The presence of microscopic voids and defects within the UHMWPE exacerbates its tendency to undergo wear and to generate particulate debris[55] and these defects are a common feature of medical grade UHMWPE. Overall, the susceptibility to wear of this material seems to be related to the combined effects of its plastic flow and the existence within it of voids and other defects.

Plastic flow behaviour of semi-crystalline polymers, such as UHMWPE, is complex. It involves yield, as molecules in the amorphous regions are drawn past each other and into new regions of space, possibly resulting in stress-induced crystallisation and strain-hardening,[56–58] UHMWPE itself is almost incompressible during elastic processes, with a Poisson's ratio of 0.46.[59] However, whether it is quite so incompressible during plastic deformations under clinical conditions is not clear.

Using finite element analysis, the maximum effective strain that develops in the artificial knee has been shown to be below 0.12.[60] Finite element studies have generally used data generated in tension, but this recent work has shown that there is a comparable strain regime under compressive loading, and this is more clinically relevant than tensile loading.

Wear of UHMWPE in orthopaedic joints is has been the subject of an enormous volume of research, and several mechanisms have been suggested to explain how such wear comes about. Under clinical conditions the two most likely are:

(i) *Abrasive wear*: This may be either two-body (femoral head on UHMWPE surface) or three-body (where there is an additional, particulate abrasive present, such as small pieces of acrylic bone cement). Such wear generates longitudinal scratches in the polyethylene and these lead to tearing and removal of material from the interface.[31]

(ii) *Abrasive fatigue*: Here, the presence of machine markings or other irregularities in the surface of the UHMWPE act as sites of initiation. Repeated movement around these defects in response to movements of the femoral head cause incremental growth in the defect with each cycle. This causes growth of the defect as a crack running transverse to the direction of the sliding motion, and containing fibrils of polymer. Large forces are not necessary for this mechanism to operate, and examination of the surface morphology of both retrieved and *in vitro* tested specimens has shown the features expected from this mechanism.[31]

The mechanism that predominates probably depends on the details of the implant geometry. The purely abrasive mechanism has been widely supported from experimental and retrieval studies, and is known to produce high rates of wear under appropriate circumstances.[61] The wear rate of polyethylene is proportional to the roughness of the counterface (Ra) to the power 1.2, *i.e.* to $Ra^{1.2}$, and a single scratch or defect in the counterface can bring about a ten-fold increase in wear rate.[62] The opposing surface of a joint involving UHMWPE must therefore not only be as smooth as possible on implantation, it must also be hard and resistant to damage so that it maintains its good surface finish throughout the lifetime of the joint.[59] There is evidence, though, that stainless steel heads articulating with UHMWPE in association with radio-opaque bone cements containing barium sulfate or zirconia wear less well than those used with radiolucent cements containing no particulate filler.[63] The conclusion from this observation is that the bone cement causes damage to the femoral head, and that particles of radio-opaque additive are acting as the third body to reduce the polish on the head and thereby increase the rate of wear. This, in turn, suggests that abrasive wear is the predominant wear mechanism in these systems.

Clinical and Biological Effects of Wear

The effect of particulate debris is to cause osteolysis in the adjacent bone tissue and consequent bone loss around the implant site.[64] This is now regarded as one of the factors contributing to the phenomenon of aseptic loosening. Loosening following infection is readily characterised, and can be prevented by appropriately stringent antiseptic techniques, but despite the use of these, a proportion of total hip arthroplasties still fail by loosening, with no sign of infection. This is what is meant by the term 'aseptic loosening'. It seems to be associated with the accumulation of wear debris, though other factors may also be involved. Certainly, where wear does occur, a variety of adverse biological effects have been reported, as shown in Table 2.3.

It is clear that the presence of particulate debris stimulates macrophage activity. The extent of this activity and any subsequent tissue pathology depends on the amount, size, geometry and chemical composition of the particles.[61] Macrophages are known to be present in considerable numbers at the interfaces of the implanted hip joint components, even those which have not undergone loosening.[70] Whether they arise as a remnant population from the tissue repair process[71] or have been attracted to the site by the presence of the foreign body surface is not clear.[61] However, their presence seems to be associated with the production of osteolytic factors, one of which is inter-leukin. One of the interleukins, IL-1, is known to be synthesised by a variety of cell types, including macrophages, and this substance seems to have the

Table 2.3 *Examples of clinical studies that describe tissue response to particulate materials (from Anstutz et al.*[64]*)*

Tissue	Effects	Reference
Periprosthetic	Implant loosening	Agins *et al.*[65]
Not specified	Bone and tissue necrosis	Revell *et al.*[66]
Periprosthetic	Granulous lesions; fracture	Santavirta *et al.*[67]
Interface, lytic zones	Bone resorption; loosening	Schmalzried *et al.*[68]
Lytic areas	Bone lysis by histiolysis	Scott *et al.*[69]

capacity to promote bone resorption.[72] In addition, macrophages secrete the cytokine known as osteoclast activating factor, OAF, and these also mediate bone loss. Finally, by the release of oxide radicals and hydrogen peroxide, macrophages also contribute to the removal of bone.[73] The resorption of bone through macrophage activity thus proceeds by a number of separate mechanisms.

The build-up of macrophage cells in the region of an implant is well known to be associated with the presence of particulate debris. The size of this debris determines the nature of the response *in vivo*. Particles at the upper end of the range, *i.e.* of the order of hundreds of micrometres, are surrounded by and engulfed by giant cells; smaller particles, by contrast, are ingested by macrophages. This provokes the physiological response known as 'activation', causing the cell to enlarge and release inflammatory and potentially osteolytic factors.[74] There seems strong evidence that macrophages can absorb bone directly, and polyethylene wear debris has been observed to be associated with histocytic granulomas, which are seen to have invaded the femoral head and neck and replaced bone, bringing about failure in otherwise well-fixed components.[61] Overall, there seems to be a range of possible effects caused by particulate debris, which lead in turn to a spectrum of cellular responses, only some of which lead to bone resorption and consequent loosening.

Where significant bone resorption occurs, it is generally identified as loss of the medial cortex or of the alteration in the bone of the proximal femur from anisotropic trabecular bone to the more isotropic cancellous bone. This has been suggested as arising from stress shielding, *i.e.* the alteration in the loading pattern of the bone adjacent to the implant, leading to bone loss in the less well-loaded regions. This may account for early bone loss, but does not seem an adequate explanation for the widespread and often late occurrence of bone loss in cemented systems.[61] Histological studies have certainly demonstrated bone loss by osteolysis,[75,76] and proximal bone loss has been shown to correlate with both high patient activity and wear of the acetabular cup.[77,78] Hence, although aseptic loosening leads to mechanical problems in the replacement joint, it is not purely mechanical in origin. Rather, it is partly caused by the accumulation of wear debris from the articulating part of the joint, and the biological consequences of this accumulation. There is a significant volume of research into this problem underway at the moment, and one promising line of research is in the development of more wear-resistant ceramic heads for the femoral component.

Crosslinked UHMWPE

An alternative approach to the development of reduced-wear replacement joints is to employ crosslinked polyethylene, and this approach has been studied and refined in recent years. It has long been known that the presence of crosslinks within polyethylene improves the abrasion resistance of the polymer, and this has been widely exploited in industrial applications.[79] However, until recently, this technology had not been employed in orthopaedics. Now, with the growing understanding of the problems of wear debris from UHMWPE, this approach is under active consideration, and there have been several reports of studies of crosslinked polyethylenes, both clinical,[80,81] and from mechanical hip simulators.[82]

Crosslinking of UHMWPE is achieved by one of three chemical routes, namely free-radical generation using ionising radiation, reaction with peroxides and reaction with silanes. These will now be considered in detail.

The interaction of polymers with all types of ionising radiation leads to the formation of free radicals. This includes the sterilising dose of gamma rays of between 25 and 40 kGy, which promotes homolytic chain cleavage leading to the formation of a pair of free radicals. Because of their proximity, many of these simply recombine. Others, however, can exist for some considerable time within the structure, and eventually combine with another polymer molecule to form crosslinks. Although the sterilisation dose of gamma rays will promote this process, UHMWPE requires a higher dose, typically at least 50 kGy, in order to become highly crosslinked.[83] For maximum crosslink density, an even greater dose is required, typically of the order of 100–150 kGy.[80]

Organic peroxides can also be used to crosslink UHMWPE. In this process, small amounts of peroxide, typically 0.1–0.2% by mass, are added to the polymer. This is sufficient to produce a highly crosslinked structure,[7] one which shows significantly altered mechanical behaviour and improved wear properties. A variety of peroxides have been used, such as dicumyl and 2,5-dimethyl-2,5-bis(*tert*-butyl peroxide).[3] They are added to the UHMWPE prior to consolidation using either compression moulding or hot-isostatic pressing. At the elevated temperatures existing in these processes, the peroxide decomposes by a free radical mechanism and leads to the formation of the crosslinks between the polymer molecules.

Finally, silanes have been used to crosslink polyethylene, but this was high density PE rather than UHMWPE.[84] This also led to improvements in mechanical properties, as well as reduced creep and diminished wear.[85]

The wear behaviour of all of these types of crosslinked polyethylene is superior to that of the uncrosslinked polymer. For example, work by Oonishi,[80] shown in Table 2.4, demonstrated a 27% reduction in wear rate in a clinical study, though the number of patients receiving the irradiated polyethylene was much lower than the control group.

Table 2.4 *Clinical wear data on gamma-irradiated UHMWPE*[80] *(against a 28 mm alumina femoral head)*

Acetabular cup	Number of patients	Average follow-up (months)	Mean wear rate mm/year
UHMWPE	71	6	0.098
Gamma-irradiated UHMWPE	9	6	0.072

Much earlier, Grobbelaar *et al.*[78] had investigated a variety of routes to crosslinking UHMWPE for use in total hip arthroplasty. They used gamma doses of up to 800 kGy, and the presence of various atmospheres, including nitrogen, acetylene and chlorotrifluoroethylene. Of these, acetylene was found to have a marked sensitising effect and to improve the crosslink density significantly for any given combination of gamma ray dose and exposure time. These materials were not tested under clinical conditions, but by *in vitro* methods. Abrasion resistance by the sand-slurry method increased by 30% compared with the uncrosslinked polymer, whereas impact strength and elongation were also found to have decreased compared with the starting material. These latter changes are also consistent with significant increases in the crosslink density in the polymer.

Peroxide crosslinked UHMWPE has been evaluated for wear behaviour in a multi-directional hip simulator.[79] Wear tests were carried out on compression moulded acetabular cups that had 1% peroxide added to them. Wear was determined as mass loss per million cycles, and was 23.3 mg per million cycles for the uncrosslinked polymer, but only 1.1 mg per million cycles for the peroxide crosslinked polymer,[79] a reduction of over 95%.

There has been relatively little work done on silane crosslinked polyethylene, possibly because the parent polymer is only high density polyethylene, the mechanical properties of which make this material unsuitable for use in orthopaedics. Indeed, it has been shown that silane crosslinked polyethylene, when tested on a mulitidirectional pin-on-disk wear tester at 37 °C in the presence of 30% bovine serum, has only the same order of resistance to wear as uncrosslinked UHMWPE.[86] There must be doubt, therefore, whether this approach is a viable one to the improvement of wear for clinical materials.

The latest generation of UHMWPE components for use in orthopaedics are sterilised under reduced-oxygen conditions to eliminate the formation of oxygenated functional groups in the surface of the polymer. Radiation-crosslinking of these materials has been achieved using either gamma rays or electron beams, with the polymer in either the solid or molten state. For example, in 1995 the company Howmedica (now Howmedica-Stryker, Rutherford, New Jersey, USA) introduced a material with the trade name Duration™ which is oxidation-stabilised and crosslinked. This polymer is fabricated in a two-step process. The first step involves gamma irradiation of the UHMWPE in a package filled with inert gas. The second step involves heat treatment at between 37 and 50 °C for 144 hours. This latter step allows the free radicals generated in the first step to recombine in a low-oxygen environment, thereby crosslinking the polymer without the formation of potentially damaging peroxy species.[87]

Wear testing of Duration™ on a multidirectional hip simulator showed it to have much less wear than an air-irradiated control specimen of UHMWPE.[88] On the other hand, under the simpler linear motion prevalent in the knee, such a material showed no improvement in wear compared with polymer that had not undergone radiation crosslinking.[89] Thus, the requirements of the polymer in hip and knee joints have been shown to differ. The goal of all this effort, though, remains the same for both applications, namely the development of improved wear behaviour in the bearing surfaces, leading to reduction in debris, and implant failure through peri-prosthetic osteolysis. Whether the recently introduced grades of UHMWPE achieve this end in clinical service remains to be seen, and much more research is needed before this question can be answered.

Hydroxypatite-filled Polyethylene

The development of hydroxyapatite-filled polyethylene was driven by the aim of preparing a synthetic material whose modulus approximated to that of bone. Just as bone is a composite material, in which hydroxyapatite acts as the filler in a polymeric matrix, so these new materials are composites, with hydroxyapatite as filler. The matrix phase is high density polyethylene. To date, one of these composites, based on 40% volume percentage of hydroxyapatite in HDPE, has been commercialised under the trade name Hapex®, and is being sold for clinical application in middle ear implants.[90] These implants have been used since 1995 to replace diseased or damaged ossicles, and the material has been incorporated into 22 different designs and employed in several thousand patients.[87]

This type of composite has the potential for much wider application in surgery, provided that the mechancial properties of bone can be closely matched. This means not only matching the ultimate strength in order to prevent fracture, but ensuring that failure is ductile. A complete modulus match has been shown to occur at 50% filler loading,[91] but unfortunately with this amount of filler the material is no longer ductile. The limit of ductility appears to be at approximately 40%, which is why this level of filler is used in the commercial material. Nonetheless, even at this level the composite has mechanical properties similar to those of bone, and has been shown experimentally to be capable of being used in applications other than the ear, for example in reconstruction of the orbital floor within the skull.[92] Hapex® has been shown to encourage bone formation *in vivo*[93] and exposure *in vitro* to physiological solution was shown to have no effect on its mechanical properties.[94] This combination of properties suggests that Hapex® should be durable over quite long periods of time when used as an implant, and shows general promise beyond its current rather limited clinical uses for the future.

These composites have also been examined as possible spinal repair devices.

In such applications, a critical consideration is their ability to withstand fatigue. Studies of biaxial fatigue have been reported, for both in-phase[95] and out-of-phase[96] motion. Torsional fatigue was found to be especially damaging for these materials and the in-phase loading to cause earlier fatigue than the out-of-phase. Overall, these materials showed some promise for the chosen application, though there remains some way to go before fatigue properties are acceptable for use in spinal restoration.

2 Acrylic Bone Cements

Acrylic bone cements are based on poly(methyl methacrylate), PMMA, a material that was introduced to orthopaedics from dentistry in the late 1950s. In dentistry, PMMA had been used as the material for fabricating denture bases since the early 1940s. The material itself achieved commercial significance as a result of the development of an inexpensive and economically viable synthetic route to the monomer by Crawford at ICI in 1932.[97] This is shown in Scheme 2.1.

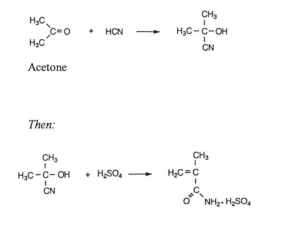

Then:

Followed by esterification with methanol:

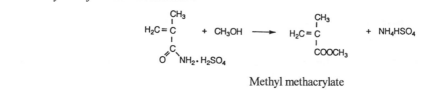

Methyl methacrylate

Scheme 2.1 *Commercial synthesis of methyl methacrylate monomer*

The pure monomer, methyl methacrylate, is a liquid with a characteristic pungent odour, boiling point 100.5 °C at atmospheric pressure, and a density of 0.936–0.940 g cm^{-3}.

For biomedical applications, including as a bone cement, methyl methacry-

Initiation:

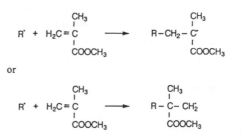

or

Propagation:

Termination:

(A) Combination:

(B) Disproportionation

Scheme 2.2 *Free radical polymerisation of methyl methacrylate*

late is made to undergo an addition polymerisation. This is typically initiated by dibenzoyl peroxide in the presence of *N,N'*-dimethyl-*p*-toluidine as accelerator. The steps in the initiation process are shown in Scheme 2.2.

Having generated the appropriate free radicals, polymerisation then continues through successive, rapid propagation reactions, leading to build up of very high molecular weight polymer chains even at low conversions. Methyl methacrylate undergoes addition polymerisation very readily, and for this reason, small amounts of inhibitor, such as hydroquinone, are added to prevent spontaneous polymerisation on storage.[98]

For biomedical uses, the dough technique is employed. Again, this has been transferred to orthopaedics from dental technology, and in this technique polymerised beads of PMMA are added to liquid monomer, typically in the ratio 3:1 by volume or 2.5:1 by mass. The polymer gradually dissolves in the monomer to form a sticky mixture. Included in the polymer beads is dibenzoyl peroxide, and this is activated by N,N'-dimethyl-p-toluidine in the liquid monomer, and thus initiates addition polymerisation. As polymerisation proceeds, the mixture rapidly becomes converted to a dough, that is, a mixture of quite high viscosity that no longer adheres to the walls of the mixing vessel. At the dough stage, the cement can be introduced to the body, either to line the freshly reamed out femur, from which the diseased head has been removed, or into the pelvic girdle. This may be achieved by hand or with the aid of a large volume syringe, and, while the cement is still dough-like, the artificial joint component, either the metal femoral pin or the UHMWPE acetabular cup, is pressed into place and held there until the cement has gelled. Completion of polymerisation takes about 15 minutes from the start of mixing.[99]

As a material for clinical use acrylic bone cement has significant drawbacks. The polymerisation of methyl methacrylate is highly exothermic, and temperatures as high as 100 °C can be reached *in vitro*, and 72 °C in the clinical situation.[100] Both the British and the US Standards for acrylic bone cements allow for a temperature rise in the laboratory of 90 °C.[101,102] Such temperature rises are potentially very damaging. Blood, for example, will coagulate at 56 °C and collagen becomes denatured at 72 °C.[103] The local temperature rise in and around a setting acrylic bone cement depends upon a number of factors, such as the volume of the cement, the reaction rate, the thermal conductivity and heat capacity of the system, the initial conditions (including temperature) and the rate of heat evolution.

While PMMA has good biocompatibility when used in total joint arthroplasty, there are problems with methyl methacrylate monomer. It is cytotoxic, and will also initiate a sudden drop in blood pressure in patients, so-called hypotension. In the past this led occasionally to post-operative cardiac arrests,[104-106] but greater awareness among anaesthetists of the problem has more or less eliminated the likelihood of hypotension proving fatal. Nonethless, the loss of free monomer from the setting bone cement still causes this effect.

The are a number of commercial bone cements available to the clinician based essentially on PMMA/methyl methacrylate chemistry (see Table 2.5). There are some variations in composition, though, for example the use of copolymers of methyl methacrylate with methyl acrylate or styrene.

The use of bone cements is very successful, especially in the hip, where approximately 90% of replacement joints last for at least 10 years. However, a significant number fail well before that, typically from aseptic loosening, with fracture of the cement mantle often being a contributary factor.[107,108]

Mixing of cements has a profound effect on their mechanical properties. There is strong evidence that voids in set cements act as sites of stress concentration and hence as points of crack initiation.[109] The pattern of their size and distribution, which varies with the mixing history of the cement

Table 2.5 *Commercial acrylic bone cements*

Brand name	Manufacturer
Simplex	Howmedica Stryker, Rutherford, NJ, USA
Palacos	E. Merck, Darmstadt, Germany
CMW	DePuy International, Blackpool, UK
Zimmer	Zimmer Inc., Indiana, USA
Sulfix	Sulzer Brothers Ltd, Switzerland

dough, has a major influence on the strength of the set material. Various mixing techniques have been employed, including hand mixing, centrifugation, vacuum and combined mechanical mixing.[110] In hand mixing, the liquid monomer and the polymer beads are combined in an open bowl with the aid of a spatula of some kind, with stirring for a recommended period of time to ensure thorough mixing. In centrifuged cements, the dough is still mixed in this way, but is then introduced into a syringe and centrifuged at speeds between 2300 and 4000 rpm at times of 30–180 seconds.[111] This has the effect of significantly reducing the porosity of the set cement.[112]

Vacuum mixing is another technique that has been employed. In this technique, hand mixing is followed by the the application of a reduced pressure through the use of a vacuum pump. The aim of this is draw out any bubbles of air or volatilised monomer, leading to a reduction in the concentration of voids.[113] However, while this has been shown to be effective in many cases, not all commercial vacuum systems are equally effective, and some have only marginal effect on the void number and distribution in certain brands of cement.[114] In general, though, studies have shown that vacuum mixing gives increased density, higher compressive strength and modulus and improved fatigue life.

Acrylic Bone Cements Containing Other Monomers

The chemistry of acrylic bone cements has received attention particularly in recent years. The use of higher alkyl methacrylates has been given much consideration in the literature. They give much less brittle polymers than PMMA, and this may have the advantage of leading to lower shear strains at the bone–cement interface due to the cushioning effect of the softer polymer.[115] One material of this type that has been studied extensively is the poly(ethyl methacrylate)–n-butyl methacrylate (PEMA–nBMA) cement developed by Braden.[116] This material is prepared by the dough technique, using pre-polymerised beads of PEMA, with n-butyl methacrylate monomer as the liquid. The polymerisation exotherm is in the range 50–55 °C, a considerably lower value than for conventional PMMA cements. Modulus of the resulting cement is lower than that of PMMA, and ductility is greater. These properties result in the PEMA cement having a superior fatigue life to PMMA cements, and also a higher fracture toughness. Residual monomer levels were found to be lower than for PMMA[117] and biocompatibility, as determined by soft tissue response in the rat, was found to be superior.[118] However, the material shows a tendency to creep, a property that is undesirable in a bone cement, and one that was improved by the inclusion of up 40% by mass of powdered hydroxyapatite filler.[119]

Crosslinking agents have also been incorporated into PEMA cements in attempts to reduce creep. A variety of difunctional dimethacrylate monomers, such as ethylene glycol dimethacrylate, EGDMA, and triethylene dimethacrylate, TEGDMA, have been used. Ultimate tensile strength was found to be improved in all cases, but setting time was shortened.[120] Also, in the case of EGDMA, the polymerisation exotherm increased, partially offsetting one of the potential advantages of this system. Clinical studies are underway with some of these cements, but to date, none has become the basis of a commercial bone cement.

Reinforced Bone Cements

Bone cements are brittle and, like most brittle materials, are weaker in tension than in compression. To increase tensile strength various fillers have been used, both fibrous and particulate materials.

Among the fibres that have been used as fillers are UHMWPE,[121] aramid[122] and carbon fibres.[123] However, results have been mixed. For example, the inclusion of UHMWPE fibres did not improve flexural strength or modulus, results that were attributed to poor mixing. The difficulties with mixing the filled cements led to the development of additional voids in the structure and this, coupled with poor interfacial adhesion, eliminated any potential reinforcing effects of the fibres. On the other hand, aramid fibres were found to improve the properties of cements in terms of fracture energy dissipation; carbon fibres proved to be even more effective than aramid fibres in this regard. For example, inclusion of carbon fibres at low volume fractions, *i.e.* 1–2%, gave increases in compressive, shear and tensile strengths, and in Young's modulus. Polymerisation exotherm was also reduced, as was creep. However, these cements were difficult to handle, because of their high viscosity, and it is not clear whether this approach is capable of yielding improved bone cements for practical applications. Certainly, no fibre-reinforced cement has yet been adopted for clinical use.

A variety of particulate reinforcing fillers have been studied. One such filler is particulate bovine bone derived from the cancellous region of the tibiae.[124] Modulus of the set cement increased with increasing content of bone particles up to about 40% by volume, and the material was successfully used *in vivo* in experiments on dogs.

Hydroxyapatite has also been used in PMMA bone cements.[125] This was found to improve physical properties, especially creep, and also Young's modulus and fracture toughness. However, as with other fillers, despite the relatively promising *in vitro* results, no commercial bone cements have yet appeared based on these materials.

3 Degradable Polymers

In many surgical applications the tissues need only temporary augmentation by a biomaterial, usually while regrowth of natural tissue occurs. In such circumstances, degradable polymers are used, sometimes referred to as bio-absorbable polymers. Clearly for such materials to be appropriate it is essential for the tissues to have sufficient capability for healing and regeneration.[126] Typical degradable implants are absorbable sutures, bone fixation devices, such as screws, rods and plates, tendon or ligament fixation tacks, and stents for urological or cardiovascular application.

In order for a polymer to be degradable, it must be susceptible to decomposition by the vital activity of the body. Generally, such biologically mediated degradation is augmented by pure physico-chemical degradation that does not depend on interaction with the tissues in order to take place.[127] Degradation depends on a number of physico-chemical factors, including crystallinity, chain mobility, surface-to-volume ratio, shape, impurities and porosity.[128]

Bioabsorbable polymers typically retain the strength to support the tissue into which they are placed for defined periods of time, which may vary from a few days to several months. The degradation products must be compatible with living tissue, at least at the concentrations in which they appear, so that the healing process may occur unimpeded. The particular advantage of such materials is that they do not require a subsequent surgical procedure for their removal and also that long-term implant-related complications are averted.

All bioabsorbable materials must fulfil certain medical and physical demands if they are to be satisfactorily deployed clinically. The physical demands are as follows:

(i)   high initial strength;
(ii) appropriate initial modulus;
(iii) controlled *in vivo* reduction in strength and modulus.

High initial strength is required because it is early in the life of the implant, while the healing tissue is at its weakest, that the implant must carry the maximum load. The controlled reduction in mechanical properties is essential, so that there is not catastrophic or unpredicatble failure of the material before the tissue has healed sufficiently to bear the loads required *in vivo*.

The main polymers used in medicine are aliphatic polyesters of α-hydroxy acids and their derivatives. A selective list of those polymers that have been found to be bioabsorbable appears in Table 2.6.

Of these, the most important are polyglycollic acid, PGA, and its copolymers and the isomeric polylactic acids, PLLA and PDLLA, and their copolymers. Polymers and copolymers of ε-caprolactone are also growing in importance.

Most of these polymers are thermoplastic, linear, partially crystalline or amorphous, and have definite melting and/or glass transition temperatures. The properties of the most important degradable polymers are given in Table 2.7.

Table 2.6 *Examples of bioabsorbable polymers*

Polyglycollic acid and its copolymers
Polylactic acid (both D and L isomers) and their copolymers
Poly-β-hydroxybutyrate
Poly-β-hydroxypropionate
Poly-ε-caprolactone
Poly-δ-valerolactone
Poly(methyl methacrylate-co-*N*-vinylpyrrolidone)
Polyvinyl alcohol
Polyanhydrides
Poly-*ortho*-esters
Polyphosphazenes

Table 2.7 *Properties of degradable polymers*

Polymer	Melting range (°C)	Glass transition temperature (°C)	Other comments
PGA	224–228	36	Insoluble in most solvents
PLLA	174–184	58	
PDLLA		57	

Degradable polymers and their blends can be processed by a variety of manufacturing techniques, and formed into a variety of shapes, including fibres and sheets, as well as screws, rods and plates. Fabrication is easier and more cost-effective than for metals or ceramics. On the other hand, the materials are relatively weak. Strengths for non-reinforced degradable polymers typically lie between 40 and 140 MPa, values which are well below the strength of cortical bone, whose bending strength has been determined as being in the range 180–195 MPa,[129,130] and whose shear strength is 68 MPa.[121] In order to raise the strength of degradable polymers to make them safe for fracture fixation, reinforcing fillers have been used, and these are considered in the next section. On the other hand, fibres of degradable polymers alone have been found to have sufficient strength to act as sutures, and they find widespread application for this purpose.

Although numerous of the degradable polymers have been proposed as degradable sutures, the main polymer used in practice is PGA. It is the polymer used in Dexon® sutures, and the main polymer used in Vicryl® sutures, where there is also a small proportion of PLLA. Degradable sutures have an excellent record when used to close soft tissue wounds and repair tendons, ligaments and dislocated joints.[131] Such fibres have also been used to construct gauzes to act as wound dressings designed to protect severe wounds, such as burns.

Degradable items for fracture fixation and related surgery have been widely studied, and materials based mainly on PGA and PLA have been found in numerous studies to give excellent results. Items such as screws,[132] rods and

plates have been evaluated clinically, not only for their biomechanical acceptability, which is satisfactory, but also for their biochemical acceptability.
Studies of the latter topic have included testing of the blood, and examination
of the patient for foreign body reaction or infection; typically, these tests reveal
no abnormalities.[133] However, it is probably important that the volume of the
implants is relatively small, and their degradation rate slow,[134] and this
prevents accumulation of acidic monomers and polymer debris in unacceptable
amounts around the degrading implants. There have been studies to suggest
that where acidic degradation products can accumulate, there are potentially
damaging effects on surrounding tissue.[135]

Although the most widespread use for these materials is as plates and screws
for temporary fixation in orthopaedics, there have been experimental studies of
their potential applications elsewhere in surgery. For example, degradable
copolymers of DL-lactide and ε-caprolactone have been used as a nerve guide
to promote functional nerve recovery in the rat.[136] In this study, experimental
animals were operated on to create a 1 cm gap in the sciatic nerve, and a
narrow-bore tube of the degradable polymer was placed to guide nerve
regrowth. The first signs of recovery were apparent after 5 weeks, but the
polymer tube had collapsed, so that it actually guided nerve growth along its
exterior. After 6 months there was a 54% recovery in motor nerve function and
a 100% recovery of sensory nerve function. The authors concluded that,
although the mechanical properties of the degradable polymer were inadequate, and it should probably be reinforced, the biological results were
extremely encouraging, and that these materials have the potential to be the
basis for novel devices for use in neuro-surgery.

Reinforced Degradable Polymers

Fibres have been widely used to reinforce degradable polymers, including
fibres of polymers that are themselves degradable. For example, polyglycollic
acid fibres have been used to reinforce poly-L-lactide plates, resulting in a
completely degradable composite whose bending modulus was increased from
4 to 6 GPa.[137] Plates of this material were found to give good results when
implanted into sheep, with no fracture during a seven-month follow-up period,
whereas the unreinforced PLLA broke within a month. Similar materials can
be made from poly-L-lactic acid, so called self-reinforced (SR) PLLA. These
composite-type materials have been found to give excellent results when
implanted into rabbits, with controlled loss of strength and normal bone
healing over a 12 week period.[138]

Also completely degradable are composites based on PLLA and PDLLA
filled with calcium metaphosphate.[139] Bending moduli of such composites were
found to lie in the range 6.4–12.8 MPa, a considerable improvement on the
unreinforced polymer. Other bioresorbable fillers explored are fibres prepared
from glasses rich in calcium and phosphorus.[140] These fillers raised the initial
physical properties considerably when included in PLLA, compared with the
unreinforced polymer. However, *in vitro* studies showed that on exposure to

Tris-buffered saline, these materials underwent an extremely rapid loss of strength, *i.e.* 37% loss after 1 week, a rate that is too great for use under clinical conditions.[121] These results are typical, and although fully degradable composite materials of this type are receiving considerable attention from researchers, they have made relatively little impact so far on clinical practice.

4 Polyurethanes

The term urethane refers to the functional group formed when an isocyanate group reacts with a hydroxyl group:

$$-N=C=O + HO- \rightarrow -NHCOO-$$

They are generally formed from difunctional ioscyanates and dihydroxy compounds, and are linear or lightly crosslinked.[1] They are effectively block copolymers, and were first suggested for use as biomaterials in 1967.[141] They have good physical properties, and also have proved to possess good haemocompatibility,[142] so have been suggested for a range of cardiovascular applications, including as small diameter vascular grafts, blood pumps and as components of an artificial heart.

Biomedical polyurethanes are generally block copolymers and consist of aromatic or aliphatic 'hard segments' and polyester, polyether or polycarbonate 'soft segments'. They typically show good haemocompatibility and polyurethane elastomers have been used in a variety of vascular devices, including heart valves,[143] heart diaphragms and vascular grafts.[144] One persistent problem with these materials is that they are subject to *in vivo* calcification,[145] a problem that can be moderated by the presence on the surface of covalently bound phosphonate groups.[146]

Of the polyurethanes, those based on polyethers have been shown to have a particularly good combination of biocompatibility and mechanical properties. However, they are susceptible to *in vivo* degradation. Such degradation starts at the methyl group α to the ether oxygen, and proceeds *via* chain cleavage, with reduction in molecular weight at the surface, and formation of surface fissures which lead in turn to loss of mechanical strength. Enzymic attack has also been shown to be possible in these materials, a consequence of the polarity of the functional groups.[147,148] Polycarbonate urethanes have recently been shown to have much better stability against oxidation than polyether urethanes in an *in vitro* study,[149] and it might be that these materials can be developed as superior components of blood-contacting devices in the future.

Although possessing inherently good haemocompatibility, polyurethanes can be improved for this application by modification to incorporate ionic groups in the polymer backbone. This approach was discovered as a result of investigations of the effect of ionic groups on platelet reactivity and coagulation action.[150] As a result, studies were carried out which demonstrated that grafting propyl sulfonate groups to the urethane nitrogen improved the blood compatibility of the polymer.[134] Similar results have been obtained using chain extenders such as 4,4'-diamino-2,2'-biphenyldisulfonic acid.[151] The use of such chain extenders, or of ionic groups, makes the molecules very hydrophilic, which in turn makes platelet deposition onto them low, yet enhances fibrinogen deposition.[152,153]

Unfortunately, despite their excellent surface chemical interaction with blood, these sulfonated polyurethanes are not suitable for use as vascular grafts in their present form, because they absorb relatively large amounts of water. However, by blending these sulfonated polyurethanes with unmodified polyurethanes, materials are obtained which absorb considerably lower amounts of water while retaining their good blood compatibility. In fact, blends showed improved interaction with blood, showing, for example, less platelet spreading when exposed to canine blood than bulk materials with the same level of sulfonation.[154] Thus, materials having acceptable physical properties and haemocomptability can be prepared from these polymers.

More recently polyurethane/liquid crystal composites have been prepared and studied as possible blood-contact devices.[155] This concept arose from the finding that biomembranes themselves are in the liquid crystal state.[156] Surface-modified polyurethanes were prepared from three different liquid crystal compounds, namely *N*-(-4-methyoxybenzylidene)-4'-ethylaniline, 4-pentyl-4'-nitrilo-biphenyl and cholesteryl oleyl carbonate, which were dissolved in THF with the polyurethane, then cast onto glass plates and the solvent allowed to evaporate. Blood compatibility was determined from a number of factors, including the platelets' adhesion to the surface and the blood clotting time. Results showed that a liquid crystal content in excess of 30% gave significant improvements in the blood compatibility of the resulting polymers, in particular by reducing the thrombogenicity of the surface.[139]

5 Hydrogels

A number of applications have been proposed for hydrogels in medicine, including as matrices for controlled drug delivery as well as in orthopaedics and ophthalmology.[157] Hydrogels themselves are three-dimensional networks of hydrophilic polymers held together, generally by covalent bonds. This prevents them from dissolving fully, yet at the same time allows them to take up considerable volumes of water, typically at least 20% of the mass of the swollen material.[158] In the dry state, hydrogels are glassy, but they become soft and swollen in the presence of water, while retaining their overall shape.

The simplest hydrogel is poly(2-hydroxyethyl methacrylate), known as polyHEMA. Lightly crosslinked versions of this polymer were first prepared by the Czech chemist Wichterle[159] and developed as soft contact lenses. The monomeric HEMA is very soluble in water, but polyHEMA, by contrast, is not water-soluble. Instead, it will take up water and swell to form a soft, elastic gel, a feature that can be controlled by including a few crosslinkable units within the molecule. Such polyHEMA hydrogels have proved to be biocompatible in a variety of locations within the body, and so have been widely used as biomaterials. Probably their most widespread application remains in ophthalmology, where they are the basis of both contact lenses and intraocular lenses.

Hydrogels and Other Polymers in Ophthalmology

The use of contact lenses to improve vision as an alternative to spectacles has increased considerably in the last twenty years or so. Lenses were originally made of individually ground glass lenses. Later, following war-time observation that splinters of PMMA in the eye caused little or no swelling or other adverse reaction, it was realised that this material was, in modern terminology, biocompatible in the eye, and it was used instead.[160] However, PMMA suffers from the same disadvantage as inorganic glasses, *i.e.* lack of oxygen permeability. The cornea has no vascular supply of its own, and therefore needs to respire directly from the atmosphere. When the oxygen supply is depleted, the Krebs cycle is disturbed, and there is an excessive accumulation of fluid, a condition known as corneal odema.

Hydrogels have better oxygen diffusion rates than PMMA or inorganic glasses. However, polymers containing sufficient water to maintain acceptable levels of oxygen diffusion tend to be soft, and this allows them to be deformed by the eyelid during blinking. There is a lower limit of stiffness in a contact lens that can be tolerated by the eye, and these high water-content materials may fall below it unless care is taken to balance crosslink density and backbone structure, so that there is limited deformation, rapid elastic recovery and good water uptake. The ideal material, particularly for extended wear, remains to be found. Meanwhile, all materials cause some damage to the eye, mainly due to inhibition of oxygen access to the cornea.

A further potential problem is that hydrogels tend to accumulate proteins from the tear fluid.[161] This may lead to contact lens allergy, and is anyway undesirable, since the adsorbed protein becomes denatured, and its distorted structure may then trigger various adverse physiological reactions from the cornea including conjunctivitis.[162] Structural rearrangements leading to denaturation occur, but these are slow processes,[163,164] and are affected by the nature of the polymer surface (chemical composition, hydrophilicity and roughness). Human serum albumin in particular has been studied for its conformational changes on adsorption onto polymers, using electron spin resonance spectroscopy,[165] and such changes have been shown to occur at a higher rate onto hydrophobic surfaces, such as PMMA, than onto hydrophilic ones. On the other hand adsorption onto the latter surfaces has still been found to occur to a significant extent, and is also accompanied by a measurable change in conformation. Research is continuing on this subject, since a complete understanding of the interaction of proteins with hydrogel polymer surfaces is essential if the performance and acceptability of contact lenses is to be improved.

The original soft contact lenses were based on the homopolymer of HEMA. They had many desirable properties and were, for example, well wetted by tear fluid and proved to more comfortable for the patient to wear than rigid

Table 2.8 *Examples of monomers used in soft contact lenses and resulting properties*

Monomers	Water content (%)	Oxygen permeability $(10^{-11}\ cm^2\ O_2\ s^{-1})$
HEMA only	38	8.4
HEMA/methacrylic acid	58	24
HEMA/methacrylic acid/N-vinylpyrrolidone	55	17
HEMA/gylceryl methacrylate	60	21
HEMA/N-vinylpyrrolidone/ 4-t-butyl-2-hydroxycyclohexyl methacrylate	66	32

PMMA lenses. However, they did suffer from disadvantages. They were mechanically fragile and showed insufficient oxygen permeability. This latter is a feature of their equilibrium water content, as shown in Table 2.8. To deal with these shortcomings, other monomers have been employed to prepare copolymers with HEMA, and these have higher water contents and correspondingly improved permeability to oxygen.

An alternative system for fabricating soft contact lenses is the silicone elastomer polydimethylsiloxane. The material has excellent optical properties, good tear resistance and very high oxygen permeability (up to $600\ 10^{-11}\ cm^2\ O_2\ s^{-1}$). However, its wettability is relatively poor, because of the hydrophobic nature of the silicone polymer. Attempts have been made to improve this, of which the most successful have been the grafting of polar polymers, such as polyethylene glycols onto the surface of the lens, thereby generating a more hydrophilic surface. Though this has achieved some success, the problem of wettability remains the major drawback of the silicone-based soft contact lenses, and they have not had the commercial success that their other desirable properties suggest they merit.

Soft contact lenses are susceptible to contamination by bacteria, which are able to adhere and may be responsible for the condiftion known as microbial keratitis, a disease that eventually leads to corneal ulceration. Three micro-organisms are commonly involved, namely *Pseudomonas aeruginosa*, *Staphylococcus aureus* and *S. epidermidis*, and they are more likely to adhere to worn contact lens surfaces that to fresh ones.[166] Pre-adsorption of lactoferrin has been found to kill such bacteria, but there may nevertheless be problems because of the release of chemokines and other inflammatory mediators, which may lead to inflammation and discomfort in the eye. The development of extended wear contact lenses requires, among other features, materials capable of selectively binding proteins that inhibit the adhesion of bacteria to the lens.[167]

Hydrogels have been used to supply drugs to the eye,[168] for example antibiotics in patients awaiting cataract surgery. Other devices have been used to deliver medication to the eye, for example to manage infections of the posterior segment, such as cytomegalovinus retinitis,[169] or to regulate the pressure in the eyeball in patients suffering from glaucoma.[144] Experimental studies have also been carried out on hydrogel devices designed to release drugs that combat conditions involving cell proliferation, such as proliferative vitreoretinopathy.[170] The use of such hydrogel delivery systems increases the bioavailability of the drug and hence allows a lower dose to be given. On the other hand, treatment is expensive and problems may arise when long term extended wear is necessary.

Intraocular lenses are another important application of hydrogels in ophthalmology. These are now used extensively to restore sight following removal of cataracts. For many years these lenses were made of PMMA, with polypropylene or poly(ethylene terephthalate) fixation hoops. However, these lenses can damage to endothelium of the cornea if they come into contact with it,[171] a result of the disruption of nutrient flow across the cornea or of mechanical irritation.[172] One approach to this problem has been to use graft coplymers of PMMA with poly(vinylpyrrolidone). These appear to have improved biocompatibility in the eye, and reduce the endothelium damage. The use of hydrogel lenses has the additional advantage that the lens is soft enough to be folded on insertion, thus reducing the trauma to the eyeball during surgery. On the other hand, their inferior mechanical properties means they are difficult to manipulate during implantation. They also have lower refractive indices, which means that a thicker lens is required to give an equivalent optical power.

So far, the conditions of the eye that have been considered are those arising from disease or other naturally occurring defect. In addition, however, there may be injuries to the eye, for example as a result of the patient's occupation, and there are biomaterial-based solutions for these under development. A common clinical procedure in response to any of these conditions, which tend to destroy the cornea, is the corneal graft, properly called a penetrating keratoplasty. Much work has been done using donor corneae but, as with most transplants, there is the problem of shortage of suitable donors. Although there may be tissue engineering solutions involving corneal repopulation using suitable cell lines,[173] polymer-based keratoprostheses seem to offer more immediate prospects for success.

These keratoprostheses are total replacements for the cornea, and the major problems with them is maintaining a clear visual window whilst, at the same time, allowing sufficient ingrowth of cells to anchor the implant in place. Another problem is that the keratoprosthesis can be extruded from the eye because of pressure inside the globe, and inadequate wound healing at the edges of the implant.[174] The polymer PMMA has been used for this application, and attempts have been made to improve its retention by coating of the implant with collagen, with some success.[175]

Recent developments in this field have been the fabrication of hybrid membranes, for example of polyHEMA grafted onto a silicone base. The resulting graft copolymer was readily colonised by epithelial cells, but suffered the disadvantage of compromised clarity.[176] Other materials, such as PTFE-based composites with large (50 μm) pore sizes, have been used, and have given acceptable success, both in terms of optical properties and absence of extrusion from the eye.[177]

Retinal detachment may be treated with a scleral buckle, designed to modify the geometry of the eyeball to enable the retina to be retained in position.[178] A variety of degradable materials have been used to fabricate these, such as collagen or stabilised fibrin, but the resulting substances have few advantages and cause numerous problems. More reliable results are obtained with non-degradable polymeric buckles, fabricated from materials such as silicones, typically in the form of sponges, or even polyethylene tubing, and these are the materials of choice for the fabrication of these devices.[162,179]

Prosthetic Ligaments and Tendons

Prosthetic materials for ligaments and tendons are another biomedical application of hydrogels that is of growing importance. Natural ligaments and tendons are dense connective tissues that consist of a protein phase (collagen and elastin) and a polysaccheride phase (proteoglycans). The strength and capability for elongation are determined by the relative amounts of these two phases, by geometrical factors and by the orientation of the individual components.[180] To mimic the physical properties of the natural materials, synthetic polymers have to be formed into some sort of structure, by processes such as weaving or braiding, and hydrogels incorporated as appropriate.[142] For example, hollow cylinders of material have been prepared using bundles of polyethylene terephthalate fibres which have been wound helically on a polyethylene former of diameter 1.5 mm. These fibres were then coated with a matrix solution of HEMA, ethylene glycol (to act as plasticiser) and ethylene dimethacrylate crosslinking agent. AIBN was used to initiate the polymerisation. The stress–strain curves differed with winding angle of the fibre bundles, but showed similarities with the stress–strain curves of natural ligaments and tendons. Dynamic mechanical properties were also similar, particularly that loss modulus decreased with increasing frequency. This behaviour is characteristic of glassy materials, and the existence of the effect within the synthetic materials was attributed to the contribution of the fibres.[142]

Finally, synthetic hydrogels have been used as replacement intervertebral discs. The natural intervertebral disc is a fibrocartilagenous structure that occupies the spaces between the vertebrae. Its function is to resist and redistribute compressive forces within the spine.[181] Unfortunately, as the spine ages, the disc loses its gel-like character, which means that compressive loads cease to be distributed evenly. In extreme cases, this leads to complete rupture of the disc, a condition that has been treated with a variety of biomaterials, including metals, metal–polymer systems and polymers alone.[142] In order to imitate the unique mechanical properties of the natural intervertebral disc, composite hydrogel systems have been suggested. These consist of polyHEMA hydrogels into which polycaprolactone has been incorporated. The presence of polycaprolactone results in materials of improved uniaxial compression properties, yet with high equilibrium water contents, *i.e.* over 25%.[182] Such materials have been studied experimentally in dogs, but have not yet been used in humans. However, they appear promising because they combine both the transport and mechanical properties needed to perform as satisfactory substitutes for natural intervertebral discs. They are also prepared from polymers whose behaviour within the body is well understood, so that gaining approval from regulatory bodies for their use is likely to be straightforward.

6 Silicone-based Implants

Silicone-based materials have a long history of application principally as mammary prostheses. Silicones themselves are polymers of silicon and oxygen, in which organic groups are attached to the silicon atoms:

$$\begin{bmatrix} & R & & R & \\ & | & & | & \\ -&Si&-O-&Si&-O- \\ & | & & | & \\ & R & & R & \end{bmatrix}_n$$

They may exhibit a variety of morphologies, including linear, lightly-branched and high crosslinked,[1] and this, together with the molecular weight, determines whether the silicone is in liquid, gel or solid form. Until recently, typical mammary prostheses were composed of a solid though deformable silicone bag, filled with an oligomeric silicone liquid. However, the use of these devices was restricted in America on April 16, 1992, by order of the US Food and Drugs Administration, who ruled that such devices would now only be available through controlled clinical studies.[183] This was highly controversial, as these devices had been used in millions of women worldwide in the previous thirty or so years.[184] However, this use was associated with concerns over a variety of adverse effects, including capsular contracture, rupture, carcinogenesis, interference with detection of cancer, and autoimmune disease.[185] In fact, a recent careful assessment of the risks involved in the use of mammary prostheses has concluded that these effects are minimal, and that there was sufficient evidence in support of this that the FDA's strict policy should revised.

Silicone within the body, though, undoubtedly leads to a reaction, typically development of fibrous tissue.[186] This is a characteristic foreign-body reaction, and proceeds with an inflammatory response. However, it does not seem to be specific to silicone and, indeed, though inflammation usually involves production of cytokines, no increase in cytokine levels was found in the bloodstream of mice into which silicone has been implanted.[187] Also, silicone gel is known to pass through the walls of the silicone bag in the form of microdroplets.[188] This is able to occur even where the bag shows no evidence of rupture. The fate of the silicone gel that escapes is not fully clear, though silicone has been detected in the lymph nodes following implantation, possibly by the action of macrophages.[189] Despite these findings, no link has been found with any systemic disease. Certainly, despite the current media attention, and the sympathetic response from the FDA, there is no suggestion that mammary implants should be removed, principally because the normal risk associated with all surgical procedures outweighs any potential risk associated with adverse reactions to silicone.[190]

7 References

1   J. W. Nicholson, *The Chemistry of Polymers*, 2nd Edition, Royal Society of Chemistry, Cambridge, UK, 1997.

2   J. A. Brydson, *Plastics Materials*, 6th Edition, Butterworth-Heinemann, Oxford, UK, 1995.

3   S. M. Kurtz, O. K. Muratoglu, M. Evans and A. A. Edidin, *Biomaterials*, 1999, **20**, 1659.

4   H. L. Wagner and J. G. Dillon, *J. Appl. Polym. Sci.*, 1988, **36**, 567.

5   B. Weightman and D. Light, *Biomaterials*, 1985, **6**, 177.

6   L. Pruitt and L. Bailey, *Polymer*, 1998, **39**, 1545.

7   R. Gul, *Improved UHMWPE for use in total joint replacement*. PhD dissertation, Massachusetts Institute of Technology, Boston, MA, 1997; Quoted in Kurtz *et al.*[3]

8   M. G. Tanner, L. A. Whiteside and S. E. White, *Clin. Orthop.*, 1995, **317**, 83.

9   J. P. Collier, M. B. Mayor, V. A. Suprenant, L. A. Dauphinais and R. E. Jensen, *Clin. Orthop.*, 1990, **261**, 107.

10 D. Swarts, R. Gsell, R. King, D. Devanathan, S. Wallace, S. Lin and W. Rohr, *Trans. 5th World Biomater. Conf.*, 1996, **2**, 196.

11 M. B. Schmidt and J. V. Hamilton, *Trans. 42nd Orthop. Res. Soc.*, 1996, **21**, 22.

12 M. Wrona, M. B. Mayor, J. P. Collier and R. E. Jensen, *Clin. Orthop.*, 1994, **299**, 92.

13 S. M. Kurtz, L. Pruitt, C. W. Jewett, R. P. Crawford, D. J. Crane and A. A. Ediden, *Biomaterials*, 1998, **19**, 1989.

14 H. McKellop, F. W. Shen, T. Ota, B. Lu, H. Wiser and E.Yu, *Trans. 43rd Soc. Biomater.*, 1997, **20**, 43.

15 M. Deng and S. W. Shelaby, *Biomaterials*, 1997, **18**, 645.

16 S. M. Kurtz, L. Pruitt, C. W. Jewett, R. P. Crawford, D. J. Crane and A. A. Edidin, *Biomaterials*, 1998, **19**, 1939.

17 J. Charnley, *Low Friction Arthroplasty of the Hip*, Springer-Verlag, Berlin, 1979.

18 M. J. Griffith, M. K. Seidenstein, D. Williams and J. Charnley, *Clin. Orthop.*, 1978, **137**, 37.

19 J. A. Davidson and G. Schwartz, *J. Biomed. Mater. Res.*, 1987, **21–A3**, 261.

20 D. F. Williams, *Concise Encyclopedia of Medical and Dental Materials*, xvii, Pergamon Press, Oxford, UK, 1990.

21 M. P. Bostrom, A. P. Bennett, C. M. Rimmac and T. M. Wright, *Clin. Orthop.*, 1994, **309**, 20.

22 P. Eyerer and Y. C. Ke, *J. Biomed. Mater. Res.*, 1984, **18**, 1137.

23 G. Scott, in *Polymers and the Environment*, Royal Society of Chemistry, Cambridge, UK, 1999, p. 54.

24 B. S. Oppenheimer, E. T. Oppenheimer, J. Danishefsky, A. P. Stout and E. F. Eirich, *Cancer Res.*, 1955, **15**, 333.

25 M. Kurth, P. Eyerer, R. Asherl, K. Dittel and U. Holz, *J. Biomater. Appl.*, 1988, **3**, 33.

26 S. P. James, *Biomaterials*, 1993, **14**, 643.
27 K. L. De Vries, R. H. Smith and B. M. Franconi, *Polymer*, 1980, **21**, 949.
28 M. S. Jahan, C. Wang, G. Scwartz and J. A. Davidson, *J. Biomed. Mater. Res.*, 1991, **25**, 1005.
29 D. H. Chenery, *Biomaterials*, 1997, **18**, 415.
30 R. S. Pascaud, W. T. Evans, P. J. J. McCullagh and D. P. FitzPatrick, *Biomaterials*, 1997, **18**, 727.
31 H. J. Nusbaum, R. M. Rose, I.L Paul, A. M. Crugnola and E. L. Radin, *J. Appl. Polym. Sci.*, 1979, **23**, 777.
32 S. M. Kurtz, C. M. Rimnac, S. Li and D. L. Bartel, *Trans. Orthop. Res. Soc.*, 1994, **19**, 20.
33 R. M. Streicher, *Radiat. Phys. Chem.*, 1988, **31**, 693.
34 P. Eyerer and Y. C. Ke, *J. Biomed. Mater. Res.*, 1984, **18**, 1137.
35 A. Bellare and R. E. Cohen, *Biomaterials*, 1996, **17**, 2325.
36 L. Pruitt and L. Bailey, *Polymer*, 1998, **39**, 1545.
37 R. W. Truss, K. S. Han, J. F. Wallace and P. H. Geil, *Polym. Eng. Sci.*, 1981, **26**, 747.
38 K. S. Han, J. F. Wallace, R. W. Rruss and P. H. Geil, *J. Macromol. Sci. Phys. B*, 1981, **19**, 313.
39 V. Premnath, W. H. Harris, M. Jasty and E. W. Merrill, *Biomaterials*, 1996, **17**, 1741.
40 M. D. Ries, K. Weaver and N. Beals, *Clin. Orthop.*, 1996, **331**, 159.
41 J. P. Collier, L. C. Sutula, B. H. Currier, J. H. Currier, R. E. Wooding, I. R. Williams, K. B. Farber and M. B. Mayor, *Clin. Orthop.*, 1996, **333**, 76.
42 S. E. White, R. D. Paxton, M. G. Tanner and L. A. Whitehouse, *Clin. Orthop.*, 1996, **331**, 164.
43 M. S. Kyi, J. Holton and G. L. Ridgeway, *J. Hosp. Infect.*, 1995, **31**, 275.
44 M. Goldman and L. Pruitt, *J. Biomed. Mater. Res.*, 1998, **40**, 378.
45 A. Wang, A. Essner, C. Stark and J. H. Dumbleton, *Biomaterials*, 1996, **17**, 865.
46 H. Oonishi, E. Tsuji and Y. Y. Kim, *J. Mater. Sci. Mater. Med.*, 1998, **7**, 393.
47 H. Oonishi, E. Tsuji and Y. Y. Kim, *J. Mater. Sci. Mater. Med.*, 1998, **9**, 575.
48 E. L. Boynton, J. P. Waddell, J. Morton and G. W. Gardiner, *Can. J. Surg.*, 1991, **34**, 599.
49 T. M. Ko, J. C. Lin and S. L. Cooper, *Biomaterials*, 1993, **14**, 657.
50 E. L. Boynton, J. Waddell, E. Meek, R. S. Labow, V. Edwards and J. P. Santerre, *J. Biomed. Mater. Res.*, 2000, **52**, 239.
51 S. M. Kurtz, L. Pruitt, C. W. Jewett, R. P. Crawford, D. J. Crane and A. Edidin, *Biomaterials*, 1998, **19**, 1989.
52 A. Wang, C. Stark and J. H. Dumbleton, *J. Biomed. Mater. Res.*, 1995, **29**, 619.
53 M. Jasty, D. D. Goetz, C. R. Bragdon *et al.*, *J. Bone Jt. Surg. Am.*, 1997, **79**, 349.
54 R. S. Pascaud, W. T. Evans, P. J. McCullagh and D. P. Fitzpatrick, *Biomaterials*, 1997, **18**, 727.
55 M. Wrona, M. B. Mayor, J. P. Collier and R. E. Jensen, *Clin. Orthop.*, 1994, **299**, 92.
56 A. Galeski, Z. Bartczak, R. E. Cohen and A. S. Argon, *Macromolecules*, 1992, **25**, 5705.
57 B. J. Lee, A. S. Argon, D. M. Parks, S. Ahzi and Z. Bartczak, *Polymer*, 1993, **34**, 3555.
58 L. Lin and A. S. Argon, *J. Mater. Sci.*, 1994, **29**, 294.
59 T. M. Wright, K. L. Gunsallus, C. M. Rimnac, D. L. Bartel and R. W. Klein, *Trans. 37th Orthop. Res. Soc.*, 1991, **16**, 248.

60 D. L. Bartel, J. J. Rawlinson, A. H. Burstein, C. S. Ranawat and W. F. Flynn Jr., *Clin. Orthop.*, 1995, **317**, 76.

61 G. H. Isaac, J. R. Atkinson, D. Dowson, P. D. Kennedy and M. R. Smith, *Eng. Med.*, 1987, **16**, 167.

62 J. R. Cooper, D. Dowson, J. Fisher and B. Jobbins, *J. Med. Eng. Technol.*, 1991, **15**, 63.

63 L. Caravia, D. Dowson, J. Fisher and B. Jobbins, *Proc. Inst. Mech. Eng.*, 1990, **204**, 65.

64 H. C. Amstutz, P. Campbell, N. Kossovsky and I. C. Clarke, *Clin. Orthop.*, 1992, **276**, 7.

65 H. J. Agins, N. W. Alcock, M. Bansal, E. A. Salvati, P. D. Wilsom, P. M. Pellici and P. G. Bullough, *J. Bone Jt. Surg.*, 1988, **70A**, 347.

66 P. A. Revell, B. Weightman, M. A. R. Freeman and B. V. Roberts, *Arch. Orthop. Trauma Surg.*, 1978, **91**, 167.

67 S. Santavirta, V. Hiokka, A. Eskola, Y. T. Konttinen, T. Paavilainen and K. Tallroth, *J. Bone Jt. Surg.*, 1990, **72B**, 980.

68 T. P. Schmalzried, L. M. Kwong, M. Jasty, R. C. Sedlacek, T. C. Haire, D. O. O'Connor, C. R. Bragdon, M. Kabo, A. Malcolm and W. H. Harris, *Clin. Orthop.*, 1991, **274**, 60.

69 W. W. Scott, L. H. Riley and H. D. Dorfman, *Am. J. Radiol.*, 1985, **144**, 977.

70 B. Levack, P. A. Revell and M. A. R. Freeman, *Acta Orthop. Scand.*, 1987, **58**, 384.

71 M. A. R. Freeman, G. W. Bradley and P. A. Revell, *J. Bone Jt. Surg.*, 1982, **64B**, 489.

72 M. Gowen, D. D. Wood, E. J. Ihrie, M. K. B. McGuire and R. C. G. Russell, *Nature*, 1983, **306**, 378.

73 G. R. Mundy, A. J. Altman, M. D. Gondeck and J. G. Bandelin, *Science*, 1977, **196**, 1109.

74 N. Kossovsky, K. Liao, B. A. Millet, D. Feng, P. A. Campbell, H. Amstutz, G. A. M. Finerman, B. J. Thomas, D. J. Kilgus, A. Cracchiolo and V. Allameh, *Clin. Orthop.*, 1991, **263**, 263.

75 N. A. Johanson, J. J. Callaghan, E. A. Salvati and R. L. Merkow, *Clin. Orthop.*, 1986, **213**, 189.

76 U. E. Pazzaglia, L. Ceciliani and P. G. Persson, *Acta Orthop. Scand.*, 1985, **104**, 164.

77 C. Hierton, G. Blomgren and U. Lindgren, *Acta Orthop. Scand.*, 1983, **54**, 384.

78 R. C. Johnston and R. D. Crowninshield, *Clin. Orthop.*, 1983, **181**, 92.

79 C. Beveridge and A. Sabiston, *Mater. Design*, 1987, **8**, 263.

80 H. Oonishi, *Orthop. Surg. Traumatol.*, 1995, **38**, 1255.

81 C. J. Grobbelaar, T. A. Du Plessis and F. Marais, *J. Bone Jt. Surg.*, 1978, **60B**, 370.

82 F. W. Shen, H. A. McKellop, R. Salovey, *J. Polym. Sci., Part B, Polym. Phys.*, 1996, **24**, 1063.

83 M. Narkis, I. Raiter, S. Shkolnik and A. Siegmann, *J. Macromol. Sci. Phys.*, 1987, **B26**, 37.

84 J. R. Atkinson and R. Z. Cicek, *Biomaterials*, 1983, **4**, 267.

85 J. R. Atkinson and R. Z. Cicek, *Biomaterials*, 1984, **5**, 326.

86 T. J. Joyce, A. Unsworth, T. M. Cartwright and D. Monk, *Trans. Br. Orthop. Res. Soc.*, 1998, 19.

87 D. C. Sun, A. Wang, C. Stark and J. H. Dumbleton, *Trans. 5th World Biomaterials Congress*, 1996, **1**, 195.

88 A. Essner, V. K. Polineni, G. Schmidig, A. Wang, C. Stark and J. H. Dumbleton, *Trans. 43rd Orthop. Res. Soc.*, 1997, **22**, 784.

89 A. Wang, V. K. Polineni, A. Essner, D. C. Sun, C. Stark and J. H. Dumbleton, *Trans. 23rd Soc. Biomater.*, 1997, **20**, 394.

90 F. J. Guild and W. Bonfield, *J. Mater. Sci.; Mater. Med.*, 1998, **9**, 497.

91 F. J. Guild and W. Bonfield, *Biomaterials*, 1993, **14**, 985.

92 R. N. Downes, S. Vardy, K. E. Tanner and W. Bonfield, *Bioceramics*, 1991, **4**, 239.

93 W. Bonfield, C. Doyle and K. E. Tanner, *Biological and Biomechanical Performance of Biomaterials*, ed. P. Cristel, A. Meunier and A. J. C. Lee, Elsevier, Amsterdam, 1986, p. 153.

94 J. Huang, L. Di Silvio, M. Wang, K. E. Tanner and W. Bonfield, *J. Mater. Sci., Mater. Med.*, 1997, **8**, 775.

95 P. T. Ton That, K. E. Tanner and W. Bonfield, *J. Biomed. Mater. Res.*, 2000, **51**, 453.

96 P. T. Ton That, K. E. Tanner and W. Bonfield, *J. Biomed. Mater. Res.*, 2000, **51**, 461.

97 J. W. C. Crawford, US Patent 2 042 458, 1932; Br. Patent 405 699.

98 R. L. Clarke, Chapter 2 in *Polymeric Dental Materials*, ed. M. Braden, R. Clarke, J. Nicholson and S. Parker, Springer-Verlag, Heidelberg, Germany, 1997.

99 E. J. Harper, *Proc. Inst. Mech. Eng. Part H*, 1998, **212**, 113.

100 B. Block, J. K. Haken and G. W. Hastings, *Clin. Orthop.*, 1970, **72**, 239.

101 British Standards Institution, BS 7253, 1990, Non-metallic materials for surgical implants, Part 1. Specification for acrylic bone cement.

102 America Society for Testing Materials, Standard 451.76, 1976, Standard specifications for acrylic bone cements.

103 R. Feith, *Acta Orthop. Scand.*, 1975, **Suppl. 161**, 1.

104 J. N. Powell, *Br. Med. J.*, 1970, 326.

105 N. H. Harris, *Br. Med. J.*, 1970, 523.

106 P. M. Frost, *Br. Med. J.*, 1970, 524.

107 T. A. Gruen, G. M. McNiece and H. C. Amstutz, *Clin. Orthop.*, 1979, **141**, 17.

108 R. N. Stauffer, *J. Bone Jt. Surg.*, 1982, **64A**, 983.

109 S. Deb, *J. Biomed. Appl.*, 1999, **14**, 16.

110 G. Lewis and G. Lewis, *J. Biomed. Mater. Res. (Appl. Biomater.)*, 1997, **38**, 155.

111 D. W. Burke, E. I. Gates and W. H. Harris, *J. Bone Jt. Surg.*, 1984, **66**, 1265.

112 M. Jasty, J. P. Davies, D. O. O'Connor, D. W. Burke, T. P. Harrigan and W. H. Harris, *Clin. Orthop.*, 1990, **259**, 122.

113 G. Lewis, J. S. Nyman and H. H. Trieu, *J. Biomed. Mater. Res.*, 1997, **38**, 221.

114 E. Fritsch, S. Rupp and N. Kaltenkirchen, *Arch. Orthop. Trauma Surg.*, 1996, **115**, 131.

115 A. S. Litsky, R. M. Rose, C. T. Rubin and E. L. Thrasher, *J. Orthop. Res.*, 1990, **8**, 623.

116 M. Braden, US Patent 4 791 150, 1988.

117 K. W. M. Davy and M. Braden, *Biomaterials*, 1991, **12**, 540.

118 P. Revell, M. Braden, B. Weightman and M. A. R. Freeman, *Clin. Mater.*, 1992, **10**, 233.

119 S. N. Khorasani, S. Deb, J. C. Beheri, M. Braden and W. Bonfield, *Bioceramics, 5, Proc. 4th Int. Symp. on Ceramics in Med.*, 1991, 301.

120 S. Deb, M. Braden and W. Bonfield, *J. Mater. Sci., Mater. Med.*, 1997, **8**, 829.

121 B. Pourdehyimi and H. D. Wagner, *J. Biomed. Mater. Res.*, 1989, **23**, 63.
122 B. Pourdehyimi, H. H. Robinson, P. Schwartz and H. D. Wagner, *Ann. Biomed. Eng.*, 1986, **14**, 277.
123 S. Saha and S. Pal, *J. Biomed. Mater. Res.*, 1986, **21**, 4486.
124 J. L. Williams and W. J. H. Johnson, *J. Biomech.*, 1989, **22**, 673.
125 J. Perek and R. M. Pilliar, *J. Mater. Sci. Mater. Med.*, 1992, **3**, 333.
126 P. Tormala, T. Pohjonen and P. Rokkanen, *Proc. Inst. Mech. Eng., Part H*, 1998, **212**, 101.
127 M. Vert, S. M. Li, G. Spenlehauer and P. Guerin, *J. Mater. Sci., Mater. Med.*, 1993, **3**, 432,
128 K. J. L. Burg and S. W. Shalaby, *J. Biomater. Sci. Polym. Ed.*, 1997, **9**, 15.
129 A. J. Tonino, C. L. Davidson, P. J. Klopper and L. A. Linclau, *J. Bone Jt. Surg.*, 1977, **58B**, 107.
130 D. T. Reilly and A. H. Burstein, *J. Biomech.*, 1975, **8**, 393.
131 R. Schedl and P. Fasol, *J. Trauma*, 1979, **19**, 189.
132 R. W. Bucholz, S. Henry and M. B. Henley, *J. Bone Jt. Surg.*, 1994, **76A**, 319.
133 T. Yamamuro, Y. Matsusue, A. Uchida, K. Shimada, E. Shimozaki and K. Kitaoka, *Int. Orthop.*, 1994, **18**, 332.
134 I. Grizzi, H. Garreau, S. Li and M. Vert, *Biomaterials*, 1995, **16**, 305.
135 Y. Matsusue, S. Hanafusa, T. Yamamuro, Y. S. Likinami and Y. Ikada, *Clin. Orthop. Rel. Res.*, 1995, **317**, 246.
136 M. F. Meek, W. F. A. den Dunnen, H. L. Bartels, A. J. Pennings, P. H. Robinson and J. M. Schakenraad, *Cells Mater.*, 1997, **7**, 53.
137 M. Vert, R. Chalbot, J. Leray and P. Christel, US Patent 4 279 249, 1981.
138 M. J. Manninen, U. Paivarinta, H. Patiala, P. Rokkanen, R. Taurio, M. Tamminmaki and P. Tormala, *J. Mater. Sci.; Mater. Med.*, 1992, **3**, 245.
139 B. S. Kelley, R. L. Dunne, G. C. Battistone, R. A. Casper, J. W. Vincent and R. A. Boronski, *Proc. 12th Annual Meeting Soc. Biomater.*, 1986, 167.
140 T. C. Lin, *Proc. 12th Annual Meeting Soc. Biomater.*, 1986, 165.
141 J. H. Boretos and W. S. Pierce, *Science*, 1967, **158**, 1481.
142 M. D. Leah and S. L. Cooper, *Polyurethanes in Medicine*, CRC Press, Boca Raton, FL, 1986.
143 T. G. Mackay, D. J. Wheatley, G. M. Bernacca, A. C. Fisher and C. S. Hindle, *Biomaterials*, 1996, **17**, 1857.
144 A. Edwards, R. J. Carson, M. Szycher and S. Bowald, *J. Biomater. Appl.*, 1998, **13**, 23.
145 G. M. Bernacca, T. G. Mackay, R. Wilkinson and D. J. Wheatley, *Biomaterials*, 1995, **16**, 279.
146 R. R. Joshi, J. R. Frautschi, J. Philips and R. J. Levy, *J. Appl. Biomater.*, 1994, **5**, 65.
147 D. J. Williams, *J. Bioeng.*, 1997, 1279.
148 S. K. Phua and J. M. Anderson, *J. Biomed. Mater. Res.*, 1987, **21**, 231.
149 M. C. Tanzi, S. Fare and P. Petrini, *J. Biomater. Appl.*, 2000, **14**, 325.
150 C. D. Ebert, E. S. Lee, J. Deneris and S. W. Kim, *ACS Adv. Chem. Ser.* 1982, **199**, 161.
151 J. P. Santerre, N. Vanderkamp and J. L. Brash, *Trans. Soc. Biomater.*, 1989, **12**, 113.
152 T. G. Grasel and S. L. Cooper, *J. Biomed. Mater. Res.*, 1989, **23**, 311
153 K. K. S. Hwang, T. A. Speckhard and S. L. Cooper, *J. Macromol. Sci. Phys.*, 1984, **B23**, 153.

154 R. W. Hergenrother and S. L. Cooper, *J. Mater. Sci.; Mater. Med.*, 1992, **3**, 313.
155 C. Zhou and Z. Yi, *Biomaterials*, 1999, **20**, 2093.
156 Y. Xie, J. Lui and Z. Ouyang, *Prog. Nat. Sci.*, 1995, **5**, 63.
157 N. A. Peppas, *Hydrogels in Medicine and Pharmacy*, Vols. I and II, CRC Press, Boca Raton, FL, 1987.
158 L. Ambrosio, R. De Santis and L. Nicolais, *Proc. Inst. Mech. Eng., Part H*, 1998, **212**, 93.
159 O. Wichterle and S. Lim, *Nature*, 1960, **185**, 117.
160 B. J. Tighe, *Chem. Br.*, 1992, **28**, 241.
161 M. F. Refojo and F. J. Holly, *Contact Lens J.*, 1977, **3**, 23.
162 M. R. Allansmith, D. R. Korb, J. V. Greiner, A. S. Henriguez, M. A. Simon and V. M. Finnemore, *Am. J. Ophthalmol.*, 1977, **83**, 697.
163 E. J. Castillo, J. L. Koenig and J. M. Anderson, *Biomaterials*, 1986, **7**, 89.
164 Q. Garrett and B. K. Milthorpe, *Invest. Ophthalmol. Vis. Sci.*, 1996, **37**, 2594.
165 Q. Garrett, H. J. Griesser, B. K. Milthorpe and R. W. Garrett, *Biomaterials*, 1999, **20**, 1345.
166 T. J. Williams, M. D. P. Willcox and R. P. Schnieder, *Optom. Vision Sci.*, 1998, **75**, 266.
167 R. L. Taylor, M. D. P. Willcox, T. J. Williams and J. Verran, *Optom. Vision Sci.*, 1998, **75**, 23.
168 M. J. Colthurst, R. L. Williams, P. S. Hescott and I. Grierson, *Biomaterials*, 2000, **21**, 649.
169 D. C. Musch, D. F. Martin, J. F. Gordon, M. D. Davis and B. D. Kupperman, *New Eng. J. Med.*, 1997, **337**, 83.
170 J. C. Pastor, *Surv. Ophthalmol.*, 1998, **43**, 3.
171 D. Baker and B. J. Tighe, *Contact Lens J.*, 1981, **10**, 3.
172 T. P. Werblin, R. L. Peiffer, P. S. Binder, B. E. McCarey and A. S. Patel, *Refract, Corneal Surg.*, 1992, **8**, 12.
173 M. S. Lehrer, T. T. Sun and R. M. Lavker, *J. Cell Sci.*, 1998, **111**, 2867.
174 L. J. Girard, *Cornea*, 1983, **2**, 207.
175 S. M. Kirkham and M. E. Dangel, *Ophthalmic Surg.*, 1991, **22**, 455.
176 S. D. Lee, G. H. Hsiue, C. Y. Kao and P. C. T. Chang, *Biomaterials*, 1996, **17**, 587.
177 J. M. Legeais, G. Renard, J. M. Parel, O. Serdarevic, M. Mei Mui and Y. Pouliquen, *Exp. Eye Res.*, 1994, **58**, 41.
178 C. L. Schepens and F. Acosta, *Surv. Ophthalmol.*, 1991, **35**, 447.
179 H. D. Remulla, P. A. D. Rubin, J. W. Shore, F. C. Sutula, D. J. Townsend, J. J. Woog and K. V. Jahrling, *Ophthamology*, 1995, **102**, 586.
180 J. Kastelic, A. Galetski and E. Bear, *Conn. Tissue Res.*, 1978, **6**, 11.
181 R. H. Rothman and F. A. Simeone, *The Spine*, 3rd Edition, Vol. 1, W. B. Saunders Co, Philadelphia, PA, 1992.
182 L. Ambrosio, P. A. Netti, S. Iannace, S. J. Huang and L. Nicolais, *J. Mater. Sci.; Mater. Med.*, 1996, **7**, 525.
183 D. A. Kessler, *New Eng. J. Med.*, 1992, **326**, 1713.
184 P. C. Gerszten, *Biomaterials*, 1999, **20**, 1063.
185 B. G. Silverman, S. L. Brown, R. A. Bright, R. G. Kaczmarek, J. B. Arrowsmith-Lowe and D. A. Kessler, *Ann. Int. Med.*, 1996, **124**, 744.
186 J. C. Fisher, *New Eng. J. Med.*, 1992, **326**, 1696.
187 N. R. Rose, *Arthritis Rheum.*, 1996, **39**, 1697.
188 W. H. Beekman, R. Feitz, P. J. van Diest and J. J. Hage, *Ann. Plast. Surg.*, 1997, **38**, 441.

189 N. S. Hardt, J. A. Emery, B. G. Steinbach *et al.*, *Int. J. Occup. Med. Toxicol.*, 1995, **4**, 127.
190 H. D. Attwood, R. Bates, J. S. Beckman, R. N. Bise *et al.*, *J. Akansas Med. Soc.*, 1994, **90**, 427; Quoted as Reference [10] in P. C. Gerszten, *Biomaterials*, 1999, **20**, 1063.

CHAPTER 3

Ceramics

1 Introduction

Ceramics are solids held together by both covalent and ionic bonds, and composed of metallic or semi-metallic elements.[1] These elements are usually combined with oxygen, but may possibly be combined instead with nitrogen, silicon or carbon. Such materials are typically formed by heat treatment, hence their name, which is derived from the Greek *keramos* (= burnt stuff). The resulting solids have characteristic properties of hardness, durability and chemical resistance. Many ceramics are based on silicates or aluminosilicates, but the class of material also includes carbides, nitrides and silicides of metals such as tungsten.

Because of their composition and structure, ceramics generally display good resistance to chemical attack; they are also resistant to the effects of elevated temperature. This is because at the atomic level they are densely packed, with small cations fitting into spaces between anions. In order to disrupt such a structure, large amounts of thermal energy are needed. Consequently, the melting temperature is high.

Many ceramics contain oxygen, and their metallic elements tend to be in a fully oxidised state. Since much deterioration of materials in their service environments is due to oxidation, the fact that ceramics are already highly oxidised means that they have significant resistance to further oxidative degradation.[1]

The chemical resistance of ceramics is not total in the environment of the body. The chemistry of the body is such that it provides a hostile environment for foreign materials[2] and this is equally true for ceramics. Some ceramics are designed to interact significantly with the body, for example Bioglass®. Others, such as alumina or zirconia, are relatively inert, despite the assaults of the body's defence mechanisms. In this chapter, we begin by looking at the types of interaction that ceramics may show within the body, then examine the medically important classes of ceramic, such as calcium phosphate cements, Bioglass® and the inert ceramics alumina and zirconia.

Table 3.1 *Examples of bioactive materials (from Thompson and Hench[4])*

Type	Trade name(s)	Bioactive behaviour
Glass	45S5 Bioglass®	Class A
Glass-ceramic	Cerabone® Cerevital®	Class B
Ceramic	Hydroxyapatite	Class B

2 Ceramics in Medicine

A number of different types of ceramic are used in medical applications,[3] and four distinct implant–tissue responses have been identified for these materials.[4] These are as follows:

(i) *Virtually inert.* Ceramics which fall into this category are those which show little or no biological response when placed within the body. Instead, they are fixed in place entirely by mechanical interaction. Materials of this type include alumina, Al_2O_3, and zirconia, ZrO, used as femoral heads in artificial hip joints.

(ii) *Porous.* These are ceramics with large surface areas, and become fixed by biological ingrowth into their pores. A typical example is hydroxy-apatite, HA, coated onto metal stems for femoral components of artificial hips.

(iii) *Bioactive.* These promote a distinct and positive biological response from the hard tissues of the body, and include HA, bioactive glasses and glass-ceramics used in various dental and orthopaedic devices. They become fixed in place by chemical bonding between the tissues and the implant.

(iv) *Resorbable.* These ceramics are more or less soluble under the conditions prevailing within the body, and are generally removed by gradual dissolution within the body causing them to be lost from their initial location. The timing of the resorption process is critical, to ensure that there has been sufficient repair or healing of the hard tissue of the body to enable it to function satisfactorily as the resorbable ceramic is lost. The main materials of this type are the calcium phosphate cements.

A variety of bioactive materials are available commercially, and examples are given in Table 3.1.

Although early work did not differentiate between different degrees of bioactivity, it is now known that materials are in fact capable of displaying a wide range of bioactive behaviour. Two classes of bioactivity have been proposed,[5] known respectively as class A and class B. Class A bioactivity has been defined as 'the process whereby a biological surface is colonised by osteogenic stem cells free in the defect environment as a result of surgical intervention'. At the interface between the tissue and the material, a class A bioactive ceramic causes both intra- and extra-cellular responses. Materials of this type are termed 'osteoproductive'.

Class B bioactivity, by contrast, is that behaviour that causes bone to grow along a material by a bioconductive pathway. There is no intra-cellular response in the hard tissue, only an extra-cellular one. Materials showing this behaviour are termed 'osteoconductive'.

3 Calcium Phosphates

Chemistry

The chemistry of the calcium phosphate system is complicated, and bone cements based on this system display an array of compositions, as well as setting properties and crystal morphology. The attractions of calcium phosphate cements as bone cements are (a) that the chemical composition of the products approximates to hydroxyapatite, the mineral phase of the body's hard tissues; and (b) that the cements set at room temperature or body temperature generally with minimal exotherm. Setting tends to be a mixture of precipitation and crystallisation, which means that the setting is actually a result either of a sol–gel transition[6] or of entanglement of precipitated crystals, a process analogous with what happens in the setting of gypsum, (calcium sulfate dihydrate, $CaSO_4.2H_2O$).[7] In fact, cements are known whose setting is predominantly by one or other of these mechanisms; thus amorphous calcium phosphate cements set solely by a sol–gel transition[8] whereas dicalcium phosphate dihydrate, $CaHPO_4.2H_2O$, cements set entirely *via* a precipitation process.[9]

The first calcium phosphate bone cement prepared was of the crystalline type, with hydroxyapatite, $Ca_{10}(PO_4)_6(OH)_2$, as the precipitated phase.[10] This has since become the basis of the commercial material known as BoneSource®. Since this first cement, a variety of calcium phosphates, as listed in Table 3.2, have been developed as bone cements. Details of them are shown in Table 3.3.

Because calcium phosphates of this type are intended for use in the body, such cements can be fabricated to include ionic species that occur naturally in the body. These include Na^+, K^+, Mg^{2+}, Ca^{2+} and H^+ as cations and PO_4^{3-}, HPO_4^{2-}, $H_2PO_4^-$, CO_3^{2-}, HCO_3^-, SO_4^{2-}, HSO_4^- and Cl^- as anions, and all have been included in cements for experimental use within the body. Most of the systems mentioned in Table 3.3 result in one predominant calcium phosphate phase being formed as the precipitate, but more complex systems are also avilable. These include Norian's Skeletal Repair System®, which is fabricated from monocalcium phosphate, α-tricalcium phosphate and calcium carbonate, and results in both dicalcium phosphate dihydrate and carbonate–apatite precipitating.[24] Similarly, mixed calcium phosphate–calcium sulfate (gypsum) bone cements have been prepared.[44]

To prepare biomedical cements in the calcium phosphate system, attention has been paid mainly to monocalcium phosphate, dicalcium phosphate

Table 3.2 *Calcium phosphates used to prepare cements*

Name	Abbreviation	Formula
Monocalcium phosphate monohydrate	MCPM	$Ca(H_2PO_4)_2.H_2O$
α-Tricalcium phosphate	α-TCP	$Ca_3(PO_4)_2$
β-Tricalcium phosphate	β-TCP	$Ca_3(PO_4)_2$
Dicalcium phosphate	DCP	$CaHPO_4$
Dicalcium phosphate dihydrate	DCPD	$CaHPO_4.2H_2O$
Tetracalcium phosphate	TTCP	$Ca_4(PO_4)_2O$

Table 3.3 *Calcium phosphate bone cements (from Driessens et al.[38])*

Combination	Precipitate
MCPM–$CaKPO_4$	ACP
MCPM–$Ca_2NaK(PO_4)_2$	ACP
H_3PO_4–β-TCP	DCPD
H_3PO_4–TTCP	DCPD, HA
MCPM–$Ca(OH)_2$	DCPD, CDHA
MCPM–TTCP	DCPD
MCPM–α-TCP	DCPD
MCPM–ClA–SWH	DCPD
MCPM–$CaCO_3$	DCPD
MCPM–β-TCP	DCPD
DCPD–TTCP	HA
DCP–TTCP	HA
α-TCP	CDHA
α-TCP–$CaCO_3$	CA

dihydrate, octacalcium phosphate and precipitated hydroxyapatite.[11] In general, the solubility and precipitation behaviour of these products is quite well established, and this has assisted understanding of the setting chemistry of these cements.

One of the features of using these materials in the body is the inherent limitation on reaction temperature, *i.e.* that it cannot exceed the body temperature, 37 °C. This means that the system may not reach equilibrium and that certain phases may be unobtainable. For example, dicalcium phosphate is not usually precipitated from an aqueous solution at room temperature,[12] but is obtained by heating the dihydrate at temperatures between 120 and 170 °C. By the same token, dicalcium phosphate is not usually hydrated below 50 °C,[13] so that should it be formed in a particular biomedical cement at body temperature it will not undergo subsequent hydration.

The overall solution behaviour of the calcium phosphate system may be influenced by other, less well characterised calcium compounds. For example, it is known that the first crystalline solid to be precipitated from weakly acidic or neutral supersaturated solutions of calcium phosphate is so-called calcium-deficient hydroxyapatite, CDHA, $Ca_9(HPO_4)(PO_4)_5OH$. This compound has

Table 3.4 *Calcium phosphates formed by precipitation between room and body temperature*

Substance	Ca:P ratio	pH range
MCPM	0.5	0.0–2.0
DCPD	1.00	2.0–6.0
OCP	1.33	5.5–7.0
CDHA	1.5	6.5–9.5
Precipitated HA	1.67	9.5–12.0

Table 3.5 *Setting reactions of calcium phosphates*

Main product	Reactions
Precipitated HA or CDHA	(i) Hydrolysis
	(ii) Reaction of TTCP
	(iii) Reaction mixtures with Ca:P < 1.67
Octacalcium phosphate, OCP	
Dicalcium phosphate dihydrate, DCPD	

well defined physical and thermodynamic properties, and is of very low solubility, K_{sp} being of the order of 10^{-85} according to some published estimates.[12]

The practical setting reactions in calcium phosphate bone cements are based on a limited number of crystalline compounds, whose formation is partly limited by their pH stability ranges.[14] These are shown in Table 3.4.

The compounds listed in Table 3.4 are the only crystalline phases that can be formed at ambient or body temperature as the result of a setting reaction between calcium phosphates in aqueous solution. Substances such as TTCP, α-TCP, β-TCP or DCP cannot be obtained as intermediates in any setting reaction.

The setting reactions of these cements fall into various groups, according to the number and type of calcium phosphates used in the powder mixture, and according to the main product formed. These are shown in summary in Table 3.5, and discussed in more detail in the sections that follow.

Hydrolysis

Because precipitated HA is the least soluble calcium phosphate phase above pH 4.2, any other calcium phosphate above this pH will dissolve and reprecipitate as HA. If this process is rapid and occurs throughout the system, cement formation will occur solely due to this hydrolysis reaction. Typically, the calcium phosphates used in cements that set by this process have Ca:P ratios lower than that of precipitated HA. This means that the precipitation of HA can only occur with the release of phosphoric acid. For example, for monocalcium phosphate monohydrate, MCPM, the reaction may be represented by:

$$5Ca(H_2PO_4)_2.H_2O \rightarrow Ca_5(PO_4)_3 + 7H_3PO_4 + 4H_2O$$

For calcium phosphate dihydrate, DCPD, it can be represented by:

$$5Ca(H_2PO_4)_2.2H_2O \rightarrow Ca_5(PO_4)_3 + 2H_3PO_4 + 9H_2O$$

The formation of phosphoric acid results in a lowering of the pH of the mixture with time. Thus, as setting proceeds, the reaction becomes slower and, in order to be driven to completion, would need the removal of the H_3PO_4 by-product. This limits the suitability of these reactions for clinical application. The addition of a basic compound to the mixture, such as NaOH, to neutralise the acid as it is formed cannot be done, because it causes additional problems due to the high initial pH value. Generally, such high pH values are avoided because in the body they cause cell death and cytotoxicity.[15]

Reactions between TTCP and Other Calcium Phosphates

Tetracalcium phosphate, TTCP, is the only calcium phosphate with a Ca:P ratio above that of precipitated HA. Because of this, it can be combined with one or more of the other calcium phosphates to obtain mixtures having an HA or CDHA stoichiometry without the formation of an acidic or basic by-product. In principle, any calcium phosphate more acidic than precipitated HA can react directly with TTCP to form HA or CDHA. For example, consider the following reactions of TTCP with MCPM:

$$7Ca_4(PO_4)O + 2Ca(H_2PO_4)_2.H_2O \rightarrow 6Ca_5(PO_4)_3OH + 3H_2O$$
$$2Ca_4(PO_4)O + Ca(H_2PO_4)_2.H_2O \rightarrow Ca_9(PO_4)_3OH + 2H_2O$$

Such reactions proceed satisfactorily at body temperature, and combine good setting and hardening properties, making the cements acceptable for biomedical application. By contrast, octacalcium phosphate and both α- and β-TCP have very low supersaturation levels in combination with TTCP, making the setting reactions too slow for practical bone cements.[16]

As is apparent, the equations above represent only the balance of reactants and products, not the primary chemical reactions. Typically these setting reactions proceed *via* the formation of DCPD, the kinetically favoured phase in this system.[17,18] However, whether the final product is precipitated HA or CDHA is controlled by the initial system stoichiometry, as indicated.

In fact, the most widely studied combinations are of TTCP with either DPCD or DPC. The former is the basis of the commercial bone repair materials Bone Source®, which is used in craniofacial surgery. The setting reactions in this material may also proceed to precipitated HA or CDHA, as can reaction of TTCP with DPC, depending on the initial ratio of starting materials. Cements formed from aqueous mixtures of these calcium phosphate materials generally show satisfactory setting and hardening behaviour for use at body temperature,[19,20] However, discussions of phase relationships based purely on thermodynamics, which are common in the literature on these materials, are often inadequate and the kinetics of various component processes of the setting reaction must also be considered if the behaviour of these materials is to be understood properly.

Reactions of Mixtures with Ca:P < 1.67

Precipitated HA can be formed as the final product using mixtures of calcium phosphates with a Ca:P ratio below 1.67 if an additional source of calcium ions is used. This means $CaCO_3$ or $Ca(OH)_2$ instead of TTCP. One such system which has been extensively studied is that of β-TCP, DPCD and $CaCO_3$ mixed as aqueous pastes,[21-23] In the setting of this cement system, crystals of precipitated HA are formed from the initial reaction of DCPD and $CaCO_3$, and these act as binders for the β-TCP particles. When the DCPD is used up, the formation of precipitated HA is controlled by the reaction between the remaining β-TCP and $CaCO_3$ particles. This latter step, though, is undesirable, since it tends to weaken the setting cement.

Another system of interest is that of α-TCP, MCPM and $CaCO_3$, which forms the basis of the Skeletal Repair System® of Norian BV mentioned previously. The fact that this is a commercial product is evidence that it has satisfactory setting and hardening characteristics. These are based on an initial reaction to form dicalcium phosphate dihydrate, followed by formation of a carbonated hydroxyapatite phase known as dahllite. This has a Ca:P ratio between 1.67 and 1.69 and a carbonate ion content similar to bone mineral.[24,25]

Reactions Forming OCP

Octacalcium phosphate is thermodynamically unstable with respect to conversion to HA. However, this conversion can only take place when both a soluble calcium salt and a source of OH^- ions are present in the reaction mixture.[26] There is also the problem that formation of OCP from aqueous mixtures of calcium phosphates is difficult because OCP is more soluble than precipitated HA. As a result, the latter phase tends to be favoured as the final product of setting.[27,28] However, there is a considerable theoretical advantage in fabricating cements containing OCP, namely that, since OCP is the precursor to the formation of HA *in vivo*, a cement containing this as the predominant crystalline phase should be very biocompatible.

OCP has been reported as being formed in a number of systems. These include from DCPD at pH 6[29] and from α-TCP.[30] Both of these processes involve simply the hydrolysis and reorganisation of a single compound. OCP has also been formed from various mixtures, such as α-TCP with DCPD,[31] DCP[32] or MCPM;[33] or from TTCP with DCPD[34] or DCP.[35]

The evidence for the formation of OCP is not fully conclusive because, for example, neither XRD nor FTIR techniques are capable of confirming its presence in a cement. Some authors have claimed that OCP has been formed on the basis that the starting mixtures had the appropriate Ca:P ratio, *i.e.* 1.33. Thermodynamic arguments have also been used, even though for OCP to be formed the necessary reactions need to be kinetically feasible as well. Overall, therefore, conclusive evidence is lacking that any practical cements really do set by the formation of OCP, at least as the predominant crystalline phase.

Reactions Forming DCPD

In the pH range 2–4.5, the dicalcium phosphates, either anhydrous or dihydrate, are the most stable calcium phosphate phases. Hence, when a basic calcium phosphate is exposed to relatively acidic conditions at ambient or body temperature, DCPD is the phase that precipitates. Cements of this type have been prepared for example from MCPM either alone or with β-TCP.

This type of cement, however, suffers from two major disadvantages for biomedical application. Firstly the requirement that the initial setting conditions be acidic is likely to lead to an inflammatory response in the tissues adjacent to the cement immediately after implantation. Secondly, it is known that DCPD can be transformed into precipitated HA over time following implantation, and this results in a loss of mechanical strength. Consequently, although they are part of the calcium phosphate system, these cements have received relatively little attention as possible bone repair materials.

Mechanical Properties of Calcium Phosphate Cements

The mechanical properties of calcium phosphate bone cements of interest are compressive and diametral tensile strength. These have usually been measured after the cements have been stored in Ringer's solution at 37 °C for 24 hours.[36] Amorphous calcium phosphates typically have low compressive strengths, generally below 15 MPa,[8] and crystalline calcium phosphates of the DCPD type have similar values.[9] By contrast, crystalline calcium phosphates of the apatite-type have quite high compressive strengths, up to 70 MPa, though the addition of accelerator tends to reduce this to around 40 MPa.[35]

In most clinical applications, calcium phosphates are placed in direct contact with human trabecular bone. It may therefore be assumed that the strength of the cement should at least match that of human trabecular bone, *i.e.* be of the order 10 MPa in compression.[37] Most calcium phosphate cements are capable of fulfilling this requirement. They have the additional advantages that their setting reactions proceed with only minimal exotherm and virtually no dimensional changes.

The preferred way of placing a calcium phosphate in clinical situations is by injection from a syringe.[38] Certainly their rheology generally makes them suitable for use in this way, though extrusion from the syringe tends to cause demixing. This has the unfortunate effect that the precise composition of a cement placed by this technique may become unknown. The addition of sodium alginate has been shown to increase the cohesion of cement pastes;[39] it has also been shown to maintain cohesion during extrusion of the cement from a syringe.[40] Consequently, this cohesion promoter makes the cement completely injectable, and setting and mechanical properties are not affected by the extrusion process.

Calcium Phosphates in the Body

The interaction of calcium phosphates with the body is complicated by the fact that body fluids such as blood and extracellular fluid contain both calcium and phosphate ions. As a result, the interaction of the implanted cement depends on the level of calcium in these body fluids, *i.e.* whether they are supersaturated, saturated or simply contain these ions at levels well below the saturation concentration ('undersaturated').[41,42] Under normal conditions of pH, *i.e.* 7.4, these fluids are supersaturated with respect to HA, saturated with respect to CDHA and CA, but undersaturated with respect to $CaCO_3$, DCP and DCPD. As a result, the latter three compounds would be expected to dissolve slowly into body fluids, even in the absence of cellular activity, whereas HA, CDHA and CA would be expected to be stable.

The quoted pH, though, is only an average, and living cells can alter the local pH in either direction. For example, the pH near macrophages and osteoclasts tends to decrease to about 5, due to formation of lactic acid. By contrast, near osteoblasts, which excrete ammonia, the pH can rise to about 8.5. This alters the pattern of solubility of the calcium phosphates. They are generally more soluble at lower pH, hence osteoclast activity can lead to dissolution of CDHA and CA, and also to slow dissolution of HA, whereas osteoblast activity can stop the dissolution of $CaCO_3$, DCP and DCPD, and also induce crystal growth of HA, CDHA and CA.[12,43] The biological behaviour of calcium phosphates thus depends upon the composition of their crystalline phases, and then in turn in a relatively straightforward way on the physical chemistry of their interaction with body fluids of differing local pH. This behaviour is now considered in detail in the sections that follow.

Gypsum-containing Calcium Phosphates

Gypsum (*i.e.* calcium sulfate dihydrate) was itself used as resorbable material in medicine until about 1960. However, resorption rates were very fast and occurred without appreciable bone growth, even where the material was modified to include HA particles.[44]

The modification of these cements to include calcium phosphate phases results in materials with improved biological properties. For example, a mixed DCPD–gypsum cement was studied as an implant in rabbits. Experimental cements were placed in the distal femoral epiphyses of rabbits, which were then sacrificed at set intervals. After 16 weeks, this cement was found to have been completely replaced by trabecular bone.[45] It was later found that the bone structure could be made more dense when excess β-TCP was added to the original cement.[46] This suggests that a fraction of the original cement, probably the gypsum component, was dissolving within the body without stimulating bone growth. Hence, even in combination with calcium phosphate, gypsum is not really acceptable as an inorganic bone cement.

Gel-like and DCPD Cements

Amorphous calcium phosphate, ACP, has a gel-like structure, and a high solubility.[47] It would therefore be expected to dissolve rapidly in body fluids. However, despite this promising behaviour, no histological study has yet been reported with calcium phosphate cements based on this material.

By contrast, the physiological behaviour of a DCPD-based cement has been reported.[48] Cylinders of this cement were implanted subcutaneously into rats to be in contact with extracellular fluid but not in direct contact with cells. The cylinders were retrieved after 8 weeks and were found to have undergone significant changes. In particularly, the proportion of phosphate had undergone a reduction, from an initial Ca:P ratio of 1.161:1 to a final Ca:P ratio of 1.403:1. In addition, the structure had become apatitic, an observation that demonstrated that the conversion of DCPD to apatite in body fluids is entirely a consequence of physical chemistry, and does not require metabolic activity of adjacent cells.

CDHA and CA Cements

Driessens *et al.*[47] also studied the *in vivo* behaviour of CDHA-based cement when implanted into rats. They used two cements, one a pure CDHA cement of Ca:P ratio 1.51:1, the other a CDHA/DCP cement with a Ca:P ratio of 1.37:1. After 8 weeks, there were virtually no changes in these Ca:P ratios, the cements had maintained their compressive and diametral strengths, and showed no change in crystal morphology. These results arise from the fact that body fluids are in equilibrium with both CDHA and DCP and hence there is no physico-chemical driving force to alter the implanted cement. Histological observations with a variety of animal species (rabbits, goats and sheep) show that these cements are gradually transformed into new bone, in a process known as osteotransduction.[37] The term osseotransduction has been coined to describe behaviour of an implanted material where there is rapid osseointegration, then slow resorption and simultaneous replacement by bone, without the formation of a gap between the implanted material and the bone.[49] This means that there is probably no loss of mechanical stability during the transformation.[37]

Histological studies have also been carried out on two commercial CA-type calcium phosphate cements. The material Norian Fracture Grout® is made from a mixture of TTCP, $CaCO_3$, H_3PO_4 crystals and dilute sodium phosphate solution. When implanted into dogs, this material was found to be osteoconductive and to show evidence of resorption at 16 weeks. Rapid bone apposition was also observed peripherally. Similarly, the Norian Skeletal

Repair System® is made from a mixture of α-TCP, $CaCO_3$, MCPM and dilute sodium phosphate solution, and good osseointegration was found for this cement when it was used to fix metal femoral prostheses.

By contrast with the gel-like and DCPD cements, cellular activity is necessary for osteotransduction of CDHA and CA-type bone cements.[37] Addition of more soluble components such as DCP or $CaCO_3$ has been proposed as a method of increasing the rate of osteotransduction,[50] though to date there have been no histological studies to confirm the effectiveness of this approach.

HA Cements

A number of biological studies have been reported on the behaviour of HA-type cements in animal models. For example, Friedman *et al.*[51] employed a cement based on TTCP and DCP or DCPD mixtures in the frontal sinus of cats. At 6 months, very little change was seen: the implanted material was more or less intact and there had been little or no new bone deposition. By 12 months, there was some evidence of change, with approximately 30% of the reconstructed region being filled with woven bone, and by 18 months bone ingrowth had reached an average of 63%, with 10% of the implanted material having been replaced by fibrovascular material. This kind of result, *i.e.* relatively slow bone deposition and equally slow cement resorption, is typical of HA cements.[37] In other animal models, though, change may be more rapid. For example, when an HA cement based on TTCP, DCPD and citric acid was implanted in the tibia of dogs, oseointegration was apparent after only two weeks.[52] However, such changes are still slower than those induced by CDHA or CA-type bone cements.

Apatite Chemistry

Having discussed calcium phosphates, and made reference to hydroxyaaptite, it is appropriate to consider in more detail the term 'apatite' and the nature of the compounds to which it is applied. In fact apatites are a group of compounds having similar though not identical structures, and with variable compositions. 'Apatite' is a broad description of these compounds, not a narrowly defined statement of composition. The substance hydroxyapatite belongs to this group, but it does have a well defined composition, *i.e.* $C_{10}(PO_4)_6(OH)_2$. It belongs to the hexagonal system with a space group $P6_3/m$. The unit cell contains Ca, PO_4 and OH ions closely packed together, and fully represents the apatite crystal.[53]

The ten calcium ions are arranged in two distinct environments, and are designated Ca(I) and Ca(II).[54] There are four calcium atoms in Ca(I) environments and six in Ca(II) environments. Of the Ca(I) atoms two are at levels $z = 0$ and two at $z = 0.5$. Of the Ca(II) atoms, three form a triangle at $z = 0.25$ and the remaining three occupy sites at $z = 0.75$, surrounding the OH groups positioned at the corners of the unit cell at $z = 0.25$ and $z = 0.75$. The PO_4

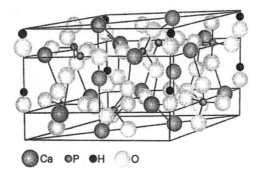

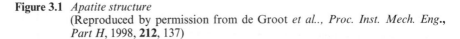
Ca P H O

Figure 3.1 *Apatite structure*
(Reproduced by permission from de Groot *et al.., Proc. Inst. Mech. Eng.,
Part H*, 1998, **212**, 137)

tetrahedra form a helix that occupies levels between $z = 0.25$ and $z = 0.75$. This
chain of PO_4 provides structural stability for the apatite crystal.

The apatite structure is readily able to vary in composition and substitutions
of one type of atom for another are common. Such substitutions may alter
various properties, such as morphology and solubility. For example, replacement of OH by F brings about a contraction in the direction of the *a*-axis as
well as reducing solubility.[55] The comparative positions of key atoms within
the Ca(II) triangles in fluorapatite and hydroxyapatite are shown in Figure 3.1.

There are a large number of other possible substitutions in the apatite
system. A few of the most important of these are shown in Table 3.6.

Biological apatites differ from pure HA in stoichiometry, composition and
crystallinity, and therefore have different physical and mechanical properties.
These species occur as the mineral phase of calcified tissue, *i.e.* enamel, dentine
and bone, and also in certain pathological conditions, such as salivary and
urinary stones. Biological apatites are typically calcium deficient and carbonate
substituted,[56] so are strictly carbonate apatites rather than hydroxyapatite.[57,58]
Bone and hydroxyapatite ceramics are compared in Table 3.7.[59]

Substitution of components in these biological apatites occurs in a coupled
manner in order to balance electrical charges, so-called Type B substitution.[60,61] This means that there is an equivalent calcium for sodium as
phosphate for carbonate modification. As a consequence of variations in these
substitutions, apatites of biological origin vary in the products they yield on
sintering. Thus, sintering enamel or dentine above 800 °C gives HA and small
amounts of β-tricalcium phosphate, whereas sintering human bone gives
mainly HA with minor quantities of CaO.

Polymer-modified Calcium Phosphate Cements

Calcium phosphate cements have been modified by the inclusion of various
types of polymer. These include natural polymers, such as cellulose deriva-

Table 3.6 *Substitutions in the apatite lattice*

Apatite	Major substituent	Lattice parameters/nm	
		a-axis	c-axis
Synthetic HA	–	944.1	688.2
Ca-deficient HA	HPO_4	943.8	688.2
Fluorapatite	F	937.5	688.0
Chlorapatite	Cl	964.6	677.1
Carbonate apatite	CO_3	954.4	685.9

Table 3.7 *Comparison of bone and hydroxyapatite ceramics (from LeGeros and LeGeros[59])*

Constituents (wt%)	Enamel	Bone	HA
Ca	36.0	24.5	39.6
P	17.7	11.5	18.5
Ca/P ratio	1.62	1.65	1.67
Na	0.5	0.7	trace
K	0.08	0.03	trace
Mg	0.44	0.55	trace
CO_3^{2-}	3.2	5.8	–
Properties			
Elastic modulus (10^6 MPa)	0.014	0.020*	0.01
Tensile strength (MPa)	70	150*	100

* Values for cortical bone.

tives,[62] sodium alginate[38] or collagen,[63] and synthetic polymers, such as poly(acrylic acid)[64] or methyl vinyl ether–maleic acid copolymer.[65] As we have seen, sodium alginate has been used to prevent demixing of the cement under the pressure of extrusion from a syringe.[38] However, its influence on setting chemistry or mechanical properties has not been studied.

On the other hand, the carboxylic acid polymers poly(acrylic acid) and methyl vinyl ether–maleic acid copolymer have been added specifically to alter the setting chemistry and mechanical properties, and these have been reported in detail.[54,55] For example, methyl vinyl ether–maleic acid copolymer was found to increase the strength of a water-setting cement based on TTCP and DCP from 50 to 70 MPa, and its flexural strength from 19.9 to 25.5 MPa. The polymer was shown by FTIR spectroscopy to undergo a chemical reaction with TTCP, forming calcium polyacrylate and HA, but leaving some residual TTCP in the cement. Reaction was slow and took 17 days to reach completion.[54]

The inclusion of polymers in these cements is still at a relatively early stage, and to date there have been few detailed studies, and none which have fully optimised materials in terms of molecular weight and viscosity of polymer solution, ratio of cement components, setting rates and final mechanical properties.

Summary of Calcium Phosphate Cements

The development of calcium phosphates as bone cements has been important because it has made available to clinicians a range of materials capable of being fabricated at room or body temperature. They are mouldable materials, and they set with little or no exotherm or dimensional change, unlike polymer-based materials. In some cases, when modified by the addition of polymer, these cements can be injectable, and this offers the possibility of considerable simplification of surgical technique.[37]

The chemistry of the calcium phosphate system is complicated, and this means that there are a wide range of compositions available, with a correspondingly large range of possible properties that can be obtained. The potential surgical applications range from orthopaedics, craniofacial reconstruction (including repair of skull defects[66]) and dentistry, in which fields these materials show a variety of generally positive biological responses.[67] There is currently a great deal of research activity on these materials, and further developments, both in their surgical applications and in their chemistry, are likely.

Calcium Phosphate Coatings

Because of their favourable biological properties, calcium phosphates have been used to coat metallic implants for use in orthopaedic surgery since the late 1980s.[68,69] Essentially there are three groups of application methods, spraying, wet methods (including electrophoresis) and sputtering.[70] They differ not only in their method of application, but also in the final form of the calcium phosphate coating deposited, as discussed below.

Spraying Techniques

The most common of all application techniques is plasma spraying, and this is used commercially to make coated implants for use in orthopaedics. In this technique, a direct current electric arc is established between two electrodes, and a stream of mixed gases is passed through this arc. The arc converts these gases into a plasma, *i.e.* an ionised, high temperature phase. At the centre of the arc, temperatures as high as 20 000 K can be achieved, but this rapidly falls with distance, so that only 6 cm from the centre, the temperature is around 2000–3000 K. A powder, typically a ceramic, can be suspended in a carrier gas that is fed into this plasma and directed at a surface while in a molten or plastic state. The substrate thus becomes coated with a layer of the powder, as shown in Figure 3.2.

Though the powder particles experience high temperatures, the substrate remains relatively cool in plasma spraying, typically in the range 100–150 °C. The technique is therefore usually considered to be a 'cold' process. The fact that the powder particles become very hot means that, for calcium phosphates, the nature of the coating deposited can vary. Calcium phosphates are thermo-

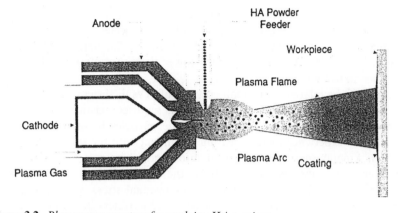

Figure 3.2 *Plasma spray system for applying HA coatings*
(Reproduced by permission from de Groot *et al.., Proc. Inst. Mech. Eng.,
Part H,* **1998, 212, 137**)

dynamically unstable at the sort of temperatures achieved at the centre of the plasma, and hence the quality and composition of the coating are very sensitive to the precise conditions employed.

In principle, the ideal situation would be for only the thin outer layer of each particle to be heated to the molten plastic state, since calcium phosphate inevitably undergoes phase changes when heated. The plastic state is necessary if the resulting coatings are to be dense and adhesive, but the fraction of each particle which achieves this state should be minimised in order retain control over the phases occurring in the final coating. In practice, this achieved by choosing an optimum balance of particle size, carrier gas, plasma speed and cooling regime.[71,72]

As an alternative to plasma spraying, low-pressure plasma spraying (LPPS) can be used. This technique, sometimes referred to as vacuum plasma spraying, employs low pressure atmospheres (50–100 mbar), which has the effect that the velocity of the particles of the powder can be increased significantly without inducing any harmful gas–metal reactions.[73,74]

The low pressure causes the plasma jet to increase considerably in size. Because of the resulting decrease in thermal gradient, substrates tend to become heated to much higher temperatures in this technique than in conventional (atmospheric pressure) plasma spraying. On the other hand, the technique has the advantage that the plasma jet velocity can be much higher, up to three times the speed of sound, due to the low pressure in the working chamber. This in turn means that the powder can also be fired at the substrate at much higher speeds, resulting in coatings with good adhesion and fewer pores.[75]

A further spray technique beginning to grow in importance as a means of applying calcium phosphate coatings is high-velocity oxy-fuel (HVOF) spraying. This is a combustion flame spray technique and employs a gun in which oxygen and a fuel, either hydrogen or propylene, are combined to generate heat and very high velocity particles. The exhaust flow can exceed the speed of sound, though the speed achieved by the injected powder particles is lower than this, typically in the range $100–800$ ms^{-1}. The flame temperature is of the order of $3000\,°C$, which is much lower than that attained by conventional plasma techniques.

Because of the relatively low temperatures of HVOF systems, it is necessary to heat the feedstock particles to give them the required plasticity. This is done by convective heat transfer, but results in the particles becoming only semi-molten. They experience further heating on impact with the substrate, as their very high kinetic energy is converted into thermal energy. On impact, the thermal energy is absorbed uniformly by the substrate, and this results in a coating which solidifies rapidly and has low residual stress. Coatings applied by the HVOF technique have higher density and hardness than most coatings applied by thermal spraying, and also have improved bonding to the substrate.[76] They also have a better crystalline structure, because the HA powder undergoes less decomposition during the coating process than in other techniques. Conversely, coatings applied by HVOF techniques tend to have the highest porosity.

Wet Techniques

A variety of techniques have been developed for low temperature deposition of calcium phosphate coatings, using suspensions or solutions of appropriate calcium phosphate salts. For example, electrophoretic deposition can be used to coat difficult substrates, such as orderly-oriented wire mesh.

In this technique, suspensions of calcium phosphate powders (3%) in isopropanol or other suitable organic liquid are used. Current is passed, and this causes the calcium phosphate powder to be deposited in the substrate at a rate and to a thickness that depend on the electric field strength. Thickness also varies with time of deposition. Following the formation of the initial coating, it is sintered at elevated temperatures ($850–950\,°C$) under high vacuum. The resulting coating consists of a mixture of calcium phosphate phases. Coatings applied in this way may suffer from cracking due to densification during sintering, and also because of thermal stresses caused by differences in thermal expansion coefficients between the metal and the calcium phosphate coating.[77,78]

Electrophoretic methods are not the only electrochemical methods that have been employed in the preparation of calcium phosphate coatings; electrodeposition has also been used.[79] In this technique, an aqueous electrolyte of calcium and phosphate ions was employed, in association with a titanium electrode. The primary calcium phosphate coating was subjected to heat treatment, an initial steam heating at $125\,°C$ for four hours, followed by vacuum calcining at $425\,°C$. The latter step caused the coating to become denser, and also improved its bonding to the substrate. Overall, this procedure was found to produce HA coatings approximately 80 μm thick, of good uniformity and consisting predominantly of small crystals of HA.

Sol–gel methods have also been used to prepare calcium phosphate coatings.[80] They have the advantage that the required coatings can be prepared with precisely defined chemical and microstructural composition. Typically, the reagents are mixed in solution, either as a colloidal suspension of inorganic particles, or as metal alkoxides or other organic precursors. The coatings are prepared at room temperature, and then fired at elevated temperatures in the region 400–1000 °C. Resulting coatings have high density and good adhesion to the substrate.

Finally in the group of wet methods that have been used comes so-called *biomimetic deposition*. This has not yet found wide application, but is based on the fact that when a scrupulously clean titanium substrate is immersed in either Hank's balanced salt solution or simulated body fluid at 37 °C for several days, a bone-like apatite layer is deposited.[81,82] The coatings prepared by this means are relatively thin, of the order of 1–5 μm, and consist of an amorphous or amorphous-crystalline structure. They have been shown to be biologically active, but despite this have not yet been used in clinical applications.

Sputter Techniques

Sputtering is the name given to the process where, under vacuum conditions, atoms or molecules are ejected from a surface as the result of bombardment by high-energy ions. The atoms or molecules then deposit on a substrate which is also placed in the vacuum chamber (Figure 3.3).

Several sputtering techniques are available, for example diode sputtering, radiofrequency or d.c. sputtering, ion-beam sputtering and reactive sputtering. However, they all suffer from the disadvantage that the deposition rate is very slow, and for this reason, they have been very used little in the preparation of calcium phosphate coatings.

Radiofrequency magnetron sputtering has been used to deposit thin (0.5–10 μm) films of HA onto titanium. X-Ray diffraction was used to show

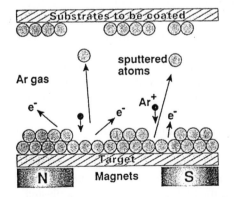

Figure 3.3 *Sputter coating system for applying HA*
(Reproduced by permission from de Groot *et al.., Proc. Inst. Mech. Eng., Part H,* 1998, **212**, 137)

that the deposited layer was amorphous or crystalline with a preferred (00*l*) crystallographic orientation with the *c* axis perpendicular to the titanium surface. The deposited film was uniform and dense, and had a Ca:P ratio between 1.5 and 2.6. It showed some solubility *in vitro* with extent of dissolution being related to the degree of crystallinity.[83-85]

Ion-assisted deposition has been used to prepare coatings. In this technique, the first step has been to prepare very thin films of a few atomic layers thickness using RF magnetron sputtering. The second stage has been to employ ion-implantation using ions such as argon, nitrogen or oxygen to form a crystalline calcium phosphate film.[86,87]

Pulsed laser deposition can be used to form thin (0.05–5 μm) HA coatings on a polished metal substrate. In this process, a target of HA is ablated by a pulsed KrF excimer laser beam of wavelength 248 nm in a low pressure environment. The ejected HA material is then deposited onto a heated substrate at temperatures of between 400 and 800 °C. The coating thus formed may be amorphous or crystalline, and shows good adhesion, particularly with high substrate temperatures, a finding that appears to be due to the existence of an oxide layer at the interface between the substrate and the film.[88,89]

Finally magnetron sputtering has also been used to prepare calcium phosphate coatings. In principle this technique offers the advantages of high deposition rate, good adhesion and the ability to coat implants of complex shape. In practice, however, difficulties remain before it can be used routinely, and further information is needed on *inter alia* the durability of the coatings and the Ca:P ratio.

Biological Interactions of Calcium Phosphate Coatings

As mentioned previously, plasma spraying is the most widely used of the application techniques, and the coatings tend to be very uniform (Figure 3.4).

Information of biological behaviour is available only for this class of coating. Such coated implants, for example, have been reported to induce more bone contact with the implant[90-93] (see Figure 3.5). They also improve implant fixation[94] and generally improve the performance when used on orthopaedic implants.[95]

HA is not the only type of calcium phosphate coating to be studied. Fluorapatite, due to its greater stability at the high temperature of the plasma, has been used to prepare plasma-spray coatings.[96] It has been shown to be more stable both *in vitro* and *in vivo* and to give more bone apposition.[80,81] However, under cyclic loading, coatings of fluorapatite on femoral stems tend to undergo delamination at the coating–substrate interface, which is undesirable.

TCP coatings have been prepared in two phases, α- and β-whitlockite. The α-form is the high temperature phase, and presumably forms when β-TCP powder is exposed to the high temperature of the plasma during spraying. The biological responses of these calcium phosphates are different. For example, less bone contact and bone remodelling has been reported with β-TCP compared with HA or α-TCP.[97] Substituting magnesium ions into the TCP

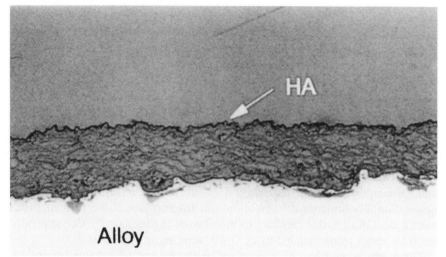

Figure 3.4 *Scanning electron microscope image of HA coating on metal alloy*
(Courtesy of Howmedica-Stryker)

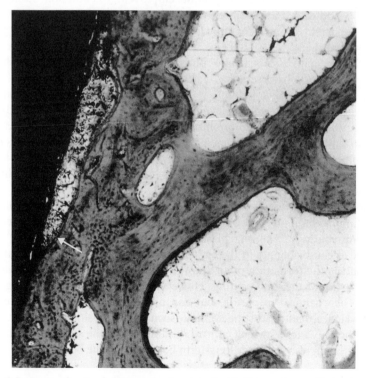

Figure 3.5 *Image showing bone growing in contact with HA coating on hip implant*
(Courtesy of Howmedica-Stryker)

lattice increases its stability[98] provided there is at least a 5% replacement.[99] On the other hand, these magnesium-substituted coatings seem to have no advantages over HA coatings because of their degradation behaviour *in vivo*.[81,100]

The effectiveness of calcium phosphate coatings varies with a number of factors. These include thickness, crystallinity, degradation behaviour, adhesion, fatigue and wear, especially third-body wear of these coatings.

In general, thickness of coatings on commercial prostheses varies from 50 to 100 μm. However, thicker coatings can be associated with fatigue failure when loaded in tension, and for this reason, de Groot has recommended that coatings should be 50–70 μm thick.[56] On the other hand, to counteract the effect of dissolution *in situ*, much thicker coatings have been recommended and Osborn[101] has suggested 200 μm as the ideal. In fact, thicker coatings unquestionably undergo delamination and fragmentation due to their brittle nature, and in time this can lead to mobility of the implant.[102] Consequently, the thicknesses recommended by de Groot seem more appropriate.

There are a number of aspects of performance of calcium phosphate coatings to be taken into account when considering their acceptability. For example, their adhesion is important, and can vary widely depending on the details of the plasma spray technique employed.[103] Adhesion strengths of between 5 and 65 MPa have been reported, with coating thickness having an influence, so that thinner coatings (50 μm) have been found to have greater adhesion that thicker ones (240 μm).[104]

The brittle nature of calcium phosphate coatings influences both their fatigue and wear characteristics. There have been several reports of cyclic loading leading to fatigue failure[92,105] and, particularly in an aqueous environment, cyclic stress can result in delamination of HA coatings. Third-body wear is also a problem, since dissolution of the relatively soluble amorphous phase may also lead to loss of crystalline particulates, and the presence of this debris has a number of undesirable consequences. There is third-body wear, leading to enhanced build up of particles, as well as loss of coating thickness. This may then lead a foreign body response[106] or to osteolysis.[107]

Dissolution behaviour and mechanical degradation can be controlled by controlling crystallinity and other aspects of the composition of the calcium phosphate coating.[108] During plasma spraying, the high temperature of the plasma can cause changes in the crystallinity of the starting material, and this may result in the formation of an amorphous or glassy phase, thus influencing the properties of the final coating. However, dissolution of the coating is not only physicochemical in nature, but can also occur as the result of cellular activity. Hence, features of the coating, such as composition and crystallinity are not the only factors which control durability, but the state of health of the nearby cells can also influence long-term behaviour in the body.

4 Porous Hydroxyapatite Implants

Synthetic porous hydroxyapatite has been considered as a bone replacement material for both load-bearing and non-load-bearing applications.[109,110] The presence of pores throughout the material gives it the potential for vascularity to develop and for bone to penetrate. A minimum pore size of 100 to 150 μm has been found to be necessary for ingrowth of fully healthy bone,[111] though good interconnectivity of the pores is also an important requirement.[112]

Varying the chemical composition of porous hydroxyapatite by inclusion of controlled amounts of different elements into the HA lattice has been used to enhance the biocompatibility of these implants.[113] This mimics the chemistry of natural HA mineral in the bone.[114] Here, as a result of the complexities of the physiological fluids surrounding the bone, there are various substitutions, for example of carbonate for phosphate and of various ions for Ca^{2+} and OH^-. By contrast, synthetic HA usually has a more closely defined composition, and readily loses carbonate ions unless sintered under carefully controlled conditions.[115]

The additional elements present in natural HA play an important part in mineral homeostasis and also in the metabolic processes of the cells in the surrounding tissue.[116] If these elements can be introduced to synthetic HA, and retained on sintering, this offers the possibility of improved biocompatibility and of beneficial biological interactions with the body. However, care needs to be taken over the level and nature of any additions. For example, very small amounts of silicon and aluminium appear to stimulate osteoblast proliferation and differentiation,[117] whereas larger amounts of aluminium suppress bone mineral formation.[104] Other important parameters are the degree of crystallinity and the phase purity of the HA present. It has been shown that variations in the precise details of the calcium phosphate strongly influence the response of cells to the implant,[118,119] partly through their influence on the solubility of the material.[120]

Porous implant materials need to be characterised for their physical properties, but conventional methods cannot be applied to them.[121] The machining of porous materials to prepare specimens for tensile, biaxial or impact testing is difficult, and gripping such specimens as part of the testing process itself causes damage, and may invalidate any results obtained.[122] Consequently, of the methods available for determining strength, only compression testing has been widely used. This has typically employed cylindrical specimens with a length to diameter ratio of 2:1. There are still inherent variations in porous samples, and testing conditions must be carefully controlled, but useful results can be obtained by this method. An additional difficulty is that materials can prove to be anisotropic, and have pores that are elongated in a particular direction, rather than being purely spherical.[109]

In one study of a porous implantable HA,[109] the material was found to act as an elastic brittle foam under uniaxial compression loading. Ultimate compressive stress and modulus was found to vary with porosity, *i.e.* with modal pore size and degree of connectivity, with the most porous specimens (apparent density 0.38 g cm^{-3}) having an ultimate compressive stress of 1 MPa, and the least porous specimens (apparent density 1.25 g cm^{-3}) an ultimate compressive stress of 11 MPa. Bone ingrowth to this material was studied and quantified in a further study.[123]

The presence of porous HA was found to accelerate fracture healing when the material was placed in cavities of rabbit femoral condyle compared with similar unfilled cavities. Implants were initially osteoconductive, and bone was observed initiating from the cavity walls and growing into the centre of the implant. This bone was shown using fluorochrome labelling to be woven, indicating that the deposition at the edge of the implant was accelerated. After this initial period, the deposited bone on the implant surfaces and within the pores became more ordered, and the deposition was consistent with osseo-inductive behaviour within the macrostructure.

Osseointegration was thus seen to be strongly influenced by the morphology of the porous HA. Moreover, ingrowth of bone altered the mechanical behavour of the material, so that after 5 weeks it no longer behaved as an elastic brittle foam but showed a more compliant behaviour, with an increased ultimate compressive stress at much higher strain levels. In other words, even after such a short time period, the material had become reinforced by ingrowth of natural bone, and its properties had become very similar to those of cancellous bone. This suggests that more porous samples of HA are actually better implant materials for repair of bone defects, because of the more rapid osseointegration they undergo, and because their lower mechanical strengths are more rapidly enhanced through the reinforcing effects of bone ingrowth. On the other hand, in order to survive surgical handling, a certain degree of inherent mechanical strength is essential, which suggests that there are practical limits to the amount of porosity permissible in an implant.[111] Consequently, for practical applications in surgery, a balance has to be struck between sufficient strength to enable the implant to be handled without damage, and sufficient porosity to allow rapid ingrowth of bone and to promote osseointegration.

5 Dense Hydroxyapatite Implants

Hydroxyapatite can also be prepared in a densified form, and there has also been considerable interest in using this material as an implant. As for the use of porous hydroxyapatite, the rationale for this interest has been that the mineral phase of bone is usually described as hydroxapatite, though as we have seen this varies somewhat in composition from chemically pure, synthetic hydroxyapatite.

Dense hydroxypatite is prepared to be relative free of porosity, and typically has a pore volume of less than 5%, and these pores in turn are typically smaller in size than 1 μm.[124] It consists of crystals of at least 200 nm in size and with physical properties shown in Table 3.8.

The preparation of dense hydroxyapatite is carried out in three stages, namely:

(i)   preparation of the apatite powder;
(ii) compaction and shaping under pressure;
(iii) sintering.

Table 3.8 *Characteristic properties of dense hydroxyapatite*

Colour	White to blue
Compressive strength/MPa	350–525
Tensile strength/MPa	35–45
Modulus of elasticity/MPa	$1.1–1.3 \times 10^4$
Impact strength/MPa	0.16–0.18

Hydroxyapatite powder can be prepared in a variety of ways, but most syntheses result in a material that is calcium-deficient, *i.e.* has a Ca/P molar ratio below 1.67. Reactions may be classified as follows:

(a) Precipitation reactions, for example the reaction of calcium nitrate and ammonium phosphate in the presence of added ammonium hydroxide:[125]

$$10Ca(NO_3)_2 + 6(NH_4)_2HPO_4 + 2NH_4OH \rightarrow Ca_{10}(PO_4)_6(OH)_2 \\ + 8NH_4NO_3 + 2HNO_3$$

Other modifications, such as the use of calcium acetate in place of calcium nitrate may be useful in preventing the incorporation of foreign anions into the apatite structure. In practice, the product is calcium-deficient, despite the apparent balance in the equation above, and this causes the formation of β-tricalcium phosphate on sintering.

(b) Hydrolysis. Hydrolysis of acidic calcium phosphates, such as dicalcium phosphate dihydrate, $CaHPO_4.2H_2O$ in alkaline solution, such as ammonium, sodium or potassium hydroxide, is a route to hydroxyapatite.

(c) Solid-state reactions. These can be brought about by mixing appropriate reagents, compressing them, and sintering above 950 °C, *e.g.* the reaction of β-tricalcium phosphate with calcium hydroxide:

$$3Ca(PO_4)_2 + 4Ca(OH)_2 \rightarrow Ca_{10}(PO_4)_6(OH)_2 + 6H_2O$$

(d) Hydrothermal reactions. Such reactions may be carried out at 2750 °C and steam pressure of 12 000 psi.[126] For example, an appropriate mixture of calcium carbonate and dicalcium phosphate can be made to yield hydroxyapatite:

$$4CaCO_3 + 6CaHPO_4 \rightarrow Ca_{10}(PO_4)_6(OH)_2 + 6H_2O + 4CO_2$$

After preparation by one of these routes, the apatite powder can be compacted and sintered to yield the dense solid hydroxyapatite. The powder may be mixed with a binder, such as starch in water or stearic acid in ethanol, though this step is not essential. The powder or slurry is then compressed or compacted at 60–80 MPa of pressure, then sintered in air at between 950 and 1300 °C, heating at a rate of about 100 °C per hour. The temperature is held at the maximum value for several hours before being lowered at the same rate as in the heating part of the process.

Alternatively, the powder may be subjected to hot-pressing, a technique in which heating and pressure are applied together. This allows densification to take place at much lower temperatures than in conventional sintering. Such reduced temperature processing prevents the formation of other calcium phosphate phases, such as α- and β-tricalcium phosphate.

Dense hydroxyapatite prepared by either of these two compaction techniques has been studied for various biomaterial applications. These include minimisation of alveolar ridge resorption following tooth extraction,[127] filling bony defects in dental or craniofacial surgery,[128,129] and as bioactive filler in polyethylene composites.[130]

Interaction of Dense Hydroxyapatite with the Body

The interaction of these materials following implantation is a function of their surface composition, as well as the local composition and pH of the body fluids in contact with them. A slightly acidic environment will cause partial dissolution of the surface, resulting in local build up of Ca^{2+} ions and various phosphate ions ($H_2PO_4^-$, HPO_4^{2-}, PO_4^{3-}). The cellular interactions depend on the surface charges that build up as these ions migrate, and this varies with initial composition. Unsintered and calcium-deficient apatite surfaces have been shown to have higher zeta potentials than ceramic or stoichiometric hydroxyapatite[131] and these are likely to affect bone formation. Proteins adsorb onto modified hydroxyapatite surfaces[132] and from such interactions eventual mineralisation may develop.

Many types of cell have been found to interact successfully with hydroxyapatite surfaces, including macrophages,[133] osteoclasts,[134] osteoblasts[135] and periodontal ligament cells.[136] Favourable cell interactions are determined in terms of cell attachment and cell proliferation, and in many cases cells appear to treat dense hydroxyapatite exactly as if it were natural bone, suggesting strong similarities in the surface chemistry of the two materials. Dense hydroxyapatite is osteoconductive, that is, it allows the formation of bone on its surface, acting as a scaffold or support for the growing and developing bone.

6 Bioactive Glasses

Bioactive glasses have numerous potential and actual applications in the repair and reconstruction of bone. The fact that certain glasses are capable of bonding to bone was first discovered by Hench and his co-workers at the University of Florida, Gainesville, in 1969, and the consequences of this finding have been the subject of extensive research in the succeeding years. Because the composition of these glasses can be finely controlled, it is possible to vary the rate of bonding to bone and also, for sufficiently reactive compositions, to extend the bonding capability to soft tissue.

Bioactive glasses are prepared by conventional glass-processing techniques, though using high purity reagents, typically of analytical reagent grade. Fusion temperatures lie typically in the range 1300–1450 °C depending on composi-

Table 3.9 *Composition of Bioglass® 45S5*

Component	Amount (%)
SiO_2	46.1
Na_2O	24.4
CaO	26.9
P_2O_5	2.6

Table 3.10 *Biological activity and composition of bioglasses*

Approximate composition range	Biological activity
(A) 35–60 mol% SiO_2; 10–50 mol% CaO; 5–40 mol% Na_2O	Bioactive, bond to bone
(B) Less than 35 mol% SiO_2	Nearly inert, the body isolates the implant by deposition of fibrous capsule
(C) Above 50 mol% SiO_2; below 10 mol% CaO; below 35 mol% Na_2O	Resorbed with 10–30 days
(D) Greater than 65 mol% SiO_2	Not technically practical. Compositions in this region have not been implanted

tion.[137] Melting needs to be carried out in non-reactive crucibles, *i.e.* those lined with platinum or platinum-based alloys, in order to prevent contamination of the melt. The most widely studied glass composition, termed Bioglass® 45S5, has the composition shown in Table 3.9.

This composition was originally selected because of the eutectic in the Na_2O–CaO–SiO_2 system at about this composition. A large number of other compositions in this region have been prepared and their properties, including bioactivity towards bone, determined.[138] The biological properties relate readily to composition, as illustrated in Table 3.10.

The precise location of the boundary compositions varies with other features of the glass; for example, the A–C boundary depends on the ratio of glass surface area to effective solution volume of the tissue, and in region C fine particle size glasses resorb more readily than bulk implants.[139]

Other compositional changes have been studied. For example, some of the CaO can be replaced by CaF_2 without altering the bone-bonding behaviour to any significant extent. The presence of fluoride does, however, affect the rate of dissolution and this, in turn, affects the precise location of the A–C boundary in the phase diagram. MgO can be substituted for CaO, and K_2O for Na_2O with little or no effect on bone bonding. Processing schedules of the glasses have been modified by additions of B_2O_3 or Al_2O_3,[140] but these do alter the interaction with bone. Alumina can inhibit bone bonding and may lead to the deposition of bone with deficiencies in the mineral phase,[141,142] so its addition is restricted to small levels, *i.e.* of the order of 1.0–1.5 wt%.

When Bioglass® and related materials were first prepared, the assumption was that some P_2O_5 was essential to promote bioactivity. However, it has since been demonstrated that phosphate-free glasses can exhibit bioactivity.[143] The role of phosphate within the glass thus appears to be only that of aiding nucleation of calcium phosphate on the surface, a function that does not critically depend on phosphate being present in the glass initially, since the surface is capable of adsorbing the necessary phosphate ions from solution.[144]

The physical properties of bioactive glasses are relatively poor. Mechanical strength is low, typically tensile bending strengths being in the region 40–60 MPa; fracture toughness is also low. On the other hand, these materials have the advantage that they undergo rapid surface reaction which leads to fast tissue bonding. Their inherent strength is therefore less critical than if they were more inert, and they are proving to be useful in applications where high strength is not required, for example as the bioactive phase in composites or as powders.

The Chemistry of the Bone–Bioactive Glass Interaction

The property of bone bonding that these glasses display is related to their chemical reactivity in body fluids. The initial reaction leads to the formation of a hydroxycarbonate apatite (HCA) layer, and bone is able to bond to this very readily. the actual bonding occurs as the result of a sequence of reactions, as follows:

When a bioactive glass is immersed in aqueous solution, three general processes take place, namely leaching, dissolution and precipitation. For these glasses, the leaching process involves alkali or alkaline earth metals, and the process may be one of ion-exchange, involving H_3O^+ entering the glass structure. The alkali or alkaline earth cations do not form part of the glass network, and so their removal from the structure is relatively easy. Because this is, partly at least, an ion-exchange process, the leaching of these cations is associated with an increase in interfacial pH to values above 7.4.

Dissolution of the glass network also occurs. This is achieved by attack of hydroxyl ions on –Si–O–Si–O–Si– units within the glass structure, and leads to breakdown of the network and the release of silica into solution in the form of silicic acid [$Si(OH)_4$]. The rate of dissolution is controlled by the composition of the glass, and decreases markedly for compositions with greater than 60 mol% SiO_2. This is a result of the much larger number of bridging oxygen atoms in glasses with high silica contents. The hydrated silica formed on the glass surface undergoes a rearrangement by condensation of neighbouring silanol groups to form a silica-rich layer.[145]

The precipitation reaction involves calcium and phosphate ions, either released from the glass or present in the adjacent body fluid, and these react to form an insoluble calcium phosphate layer which deposits on the surface. Under the conditions that occur *in vivo* this calcium phosphate forms within the condensed silica gel layer. The calcium phosphate formed in this way is initially amorphous, but it gradually incorporates carbonate anions from

solution to form a crystalline HCA structure.[146] These steps may be summarised as follows:

Stage 1: Rapid exchange of Na$^+$ or K$^+$ with H$_3$O$^+$ from solution. This stage is
usually diffusion controlled and characterised by a dependence on $t^{\frac{1}{2}}$ (where
t = time).

Stage 2: Loss of soluble silica, as a result of the reaction:

$$-Si-O-Si- + H_2O \rightarrow -Si-OH + HO-Si-$$

This is an interfacial reaction, and shows a dependence on $t^{1.0}$.

Stage 3: Condensation and repolymerisation of the SiO$_2$-rich surface layer
depleted in cations:

$$-Si-OH + HO-Si- \rightarrow -Si-O-Si- + H_2O$$

Stage 4: Migration of Ca^{2+} and PO$_4^{3-}$ groups to form precipitates of
amorphous calcium phosphate.

Stage 5: Crystallisation of the amorphous calcium phosphate by incorporation
of OH$^-$, CO$_3^{2-}$ or F$^-$ ions to form a hydroxy, carbonate, fluorapatite layer.

A number of techniques have been used to determine this sequence of
reactions, and to provide details of the steps involved. For example, Auger
electron spectroscopy (AES) has been used to study the surface layers on a
specimen of 45S5 bioactive glass that had been implanted into rat bone for one
hour.[147] Results are summarised in Figure 3.6. In this experiment the analysis
of 5 nm slices through the surface was used to build up a compositional profile
of the reaction interface. This showed that the silica-rich layer had formed to a
thickness of 1200 nm, and that the calcium phosphate layer was already 800
nm thick after only an hour. The outer layer also showed signals from carbon
and nitrogen, indicating that biological molecules had begun to deposit in this
region.

Fourier transform infrared spectroscopy (FTIR) was also used to study this
phenomenon. This technique is capable of resolving many of the important
peaks associated with key reaction products in the sequence. When FTIR was
applied to the surface of 45S5 bioactive glass stored in tris-buffer solutions at
37 °C, changes in silanol peaks were more or less complete after 20 minutes,
with a distinct new band developing at 880 cm^{-1} which was attributed to the
formation of the silica gel layer by polycondensation of the silanol groups.
After 10 minutes, a P–O bending vibration was apparent, this being associated
with the growth of the amorphous calcium phosphate layer. As time went on,
this band was replaced by two P–O modes, and these are assigned to the
crystalline apatite phase which gradually forms.

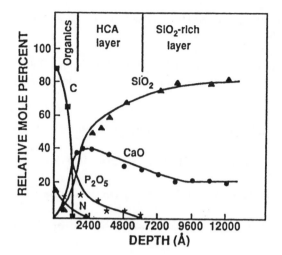

Figure 3.6 *AES analysis of the interface between Bioglass and rat bone after 1 hour's contact*[147]

The use of FTIR has allowed these surface reactions to be studied in considerable detail. For example, variations in the composition of the solution have been shown to alter the kinetics of the surface reactions. When calcium and phosphate ions were present in simulated body fluid, the rate of repolymerisation of silica was enhanced; conversely when Mg^{2+} ions were present, the rate of deposition of the amorphous calcium phosphate was slowed and the crystallisation of HCA on the surface significantly retarded. Small variations in glass composition have also been found to influence the speed of these surface reactions. Glasses with up to about 53 mol% SiO_2 show very rapid (> 2 hours) crystallisation of HCA on the surface. When studied *in vivo* such compositions not only rapidly develop a bond to bone, but they are also able to bond to collagen in soft tissues. Glasses with silica contents in the range 53–58 mol% were found to react much more slowly, taking 2–3 days to form an amorphous calcium phosphate layer and the beginnings of crystallisation to HCA. *In vivo*, these glasses will bond only to bone, not to soft tissue. Rather, where these glasses were placed in soft tissue, the response was the formation of fibrous capsule parallel to the interface and non-adherent. Compositions with SiO_2 above 60 mol% do not form a crystalline HCA layer even after four weeks in simulated body fluid. Glasses of this composition show no bioactivity and will not bond to bone.

The surface reactions considered so far are non-vital. In other words, they take place a consequence of the interaction of the glass surface with the external solution, and though of value in predicting biological interactions are not themselves anything other than non-vital *in vitro* chemical processes. Obviously, they stimulate biological reactions, and thereby are capable of bonding to vital tissues. The events that occur on the biological side of the interface are considered in the section that follows.

Tissue Response to Bioactive Glasses

Bonding to tissues occurs as the result of processes that are less well defined or understood than those that take place in the surface of the glass. However, according to Hench and Andersson, a general sequence can be determined, and is as follows:

(A) Adsorption of biological species onto the SiO_2–HCA surface layer;
(B) Action of macrophages;
(C) Attachment of stem cells;
(D) Differentiation of stem cells;
(E) Generation of matrix;
(F) Mineralisation of matrix.

Within the tissue, rapid development of HCA on the Bioglass® surface causes collagen fibrils to be incorporated. This occurs without any corresponding incorporation of cells and does not involve either enzymes or growth factors. This incorporation of collagen is similar to the process that takes place with soft-tissue bonding[148] and is a result of step (A) above.

Overall, the steps (A) to (F) take place rapidly, and by approximately 1 week mineralising bone is detectable at the interface of tissues with the more reactive bioactive glass formulations. Typically, this leads to complete bonding of the interface by four weeks, with no fibrous capsule formation. The process is always associated with a gradual immobilisation of the implant within the tissues, and so the extent of development of a sound tissue–implant interface is often determined experimentally using push-out or pull-out tests. In a typical push-out test using rabbit tibia bone as the biological tissue, reactive glasses were found to have bond strengths in the range 15–25 MPa, whereas biocompatible but unreactive glasses showed values in the range 0.5–3 MPa. Thus it is apparent that there is true bonding of bone to glass for the more reactive compositions.

Clinical Applications of Bioactive Glasses

Bioglass® 45S5 and related compositions are relatively soft, and can be drilled using standard surgical drilling procedures. The constituent elements (Si, Na, O, Ca and P) are all found in the body and loss of these elements from the implant has not been found to cause any disturbance to the tissues, at least at the concentrations that develop under clinical conditions. Tests have also shown that particles of these glasses are safe. They can, for example, be injected into the bloodstream, but have been found not to generate emboli or to have any detectable effect on blood flow.[149]

As a result of all these studies, it is now recognised that Bioglass® can be used surgically in the following forms:

1. Solid shapes;
2. Particles of varying diameters;

3. Particles combined with autologous bone particles;
4. Particles delivered *via* an injectable system.

The first successful surgical use of Bioglass® 45S5 was as a solid shape, namely the replacement for ossicles in the middle ear. This is necessary in certain patients to treat conductive hearing loss, a condition that arises due to diseases or trauma of the natural ossicles and which prevents sound being conducted from the tympanic membrane to the oval window of the inner ear. The advantage of using a bioactive material such as Bioglass® 45S5 is that there is no tendency for fibrous tissue to develop, a feature that has been associated with other middle ear implants, and one which leads to gradual loss of effectiveness of the artificial ossicle. Synthetic materials may also be extruded from the ear as part of the body's adverse reaction to them, and this feature is also absent when Bioglass® is employed.

Another successful use of solid pieces of Bioglass® is in oral surgery, where preformed cones of the material can be used to fill defects in the jaw created by the removal of teeth.[150] The effect of losing one or more teeth is that the bone in the alveolar ridge experiences a change in the loading pattern. This, in turn, causes the bone to remodel and to change in shape. Such changes often mean that the jaw can no longer comfortably support dentures. The insertion of solid Bioglass® cones has proved a useful method of overcoming this problem.

Bioglass® particles have found use in the treatment of periodontal disease.[151] This condition is caused by an infection within the gum that spreads to the bone surrounding the teeth. It occurs partly by the destruction of the soft tissues which are lost, partly exposing the bone, into which infection then spreads. This results in loss of the tooth. Treatment involves surgery to remove the diseased tissue and this may be replaced using particles of a reactive Bioglass®, which will stimulate both hard and soft tissue regeneration.

Bioglass® particles have been used with autologous bone in the recontruction of the lower jaw following tumour removal or trauma. In principle the ideal material for these reconstructions is autologous bone, for example, from the patient's own iliac crest. The difficulty is that there is only a relatively small amount of such bone available. By blending such bone with a bioactive glass, a composite material may be obtained that is safe to use, and available in sufficient quantities to complete the repair.

Other uses of Bioglass®, such as in injectable systems, are still the subject of clinical trials. However, the overall properties of these materials in terms of reactivity and speed of bonding to bone make them promising for use in a variety of surgical procedures.

7 Apatite/Wollastonite Glass-ceramics

Heat treatment of glasses leading to partial crystallisation to form ceramic phases leads to the formation of composite materials known as glass-ceramics. By controlling the size and content of crystalline phase, these glass-ceramics can be fabricated to have superior properties to the parent glass. For biomedical application, glass-ceramics have been developed that are based on apatite and wollastonite as the precipitated phases, so called A/W glass ceramics.

The initial work in this field studied the pseudo-ternary system $3CaO.P_2O_5$–$CaO.SiO_2$–$MgO.CaO.2SiO_2$.

The glass was prepared by a conventional process in which the ingredients were melted, followed by quenching. Subsequent heat-treatment of the bulk glass involving a temperature rise of 5 °C min^{-1} to a maximum of 1050 °C led to the partial crystallisation of β-wollastonite, $CaO.SiO_2$. This has a chain structure based on the silica units[152] and precipiates as a fibrous structure. Fine-grained crystallites of oxyapatite also formed.

Further studies of this general type of system have led to the development of a glass-ceramic in which the wollastonite phase is not fibrous, but is homogeneously dispersed as crystals about 50–100 nm in size. Apatite crystals are similarly dispersed. The system is based on a glass whose initial composition is as follows:

MgO	4.6 wt%
CaO	44.7 wt%
SiO$_2$	34.0 wt%
P$_2$O$_5$	6.2 wt%
CaF$_2$	0.5 wt%

Following preparation of the glass, it is densified at about 830 °C, after which first orthofluorapatite $[Ca_{10}PO_4)_6(O, F_2)]$ then wollastonite precipitate at 870 °C and 900 °C respectively. The resulting glass-ceramic is crack-free, non-porous, and dense and homogeneous.[153] X-Ray powder diffraction showed the composition to be 38% apatite, 34% wollastonite and 28% residual glassy phase, with the composition of the latter being 16.6 wt% MgO, 24.2 wt% CaO and 59.2 wt% SiO$_2$. This ceramic, named A/W after its crystalline phases, is known commercially as Cerabone® AW.[154]

A/W glass ceramics can be machined into a variety of shapes, including screws, using appropriate tools. For biomedical application, they have been formed into artificial vertebrae, spinal process spacers and iliac spacers. The physical properties that make them suitable for these uses are shown in Table 3.11. The bending strength of 215 MPa is approximately double that of dense hydroxyapatite, and even exceeds that of human cortical bone, for which the comparable figure is 160 MPa. The fracture toughness is high, and this is a function of the presence of the wollastonite. This precipitated phase inhibits straight-line propagation of cracks, causing them either to turn or to branch out. This is despite the fact that the morphology of the precipitated wollastonite is not fibrous.[155]

Mechanical properties are superior in specimens that have been kept under dry conditions. This is because in aqueous conditions the material experiences slow crack growth due to stress corrosion, behaviour which is similar to that of many other ceramics. This means that the properties when used within the

Table 3.11 *Physical properties of A/W glass ceramic*

Density (g cm^{-3})	3.07
Bending strength (MPa)	215
Compressive strength (MPa)	1080
Young's modulus (GPa)	118
Fracture toughness (MPa$^{\frac{1}{2}}$)	2.0

body will also be less good than those recorded for dry specimens, though direct data to demonstrate this point are not available.

A/W ceramics form very strong bonds to bone, so much so that fracture under tensile load tends to occur within the bone itself, rather than at the bone–ceramic interface. It is extremely bioactive, as demonstrated in experiments which were carried out by implanting granular particles into rat tibia. After 4 weeks, some 90% of the surface area was covered with newly deposited bone. By comparison, sintered hydroxyapatite particles were only 60% covered after the much longer period of 16 weeks.[156]

The bonding process begins with the formation of a very thin layer rich in calcium and phosphorus on the surface of the A/W ceramic.[157] This layer has been shown to be an apatite using X-ray microdiffraction and transmission electron microscopy has then shown that this apatite layer is the means by which the ceramic is bonded to living bone, though without the development of a distinct boundary.[158] An identical layer of apatite also forms when A/W glass-ceramic is exposed to simulated body fluid with an ionic composition similar to that of human blood plasma, and this has been shown to be a carbonate enriched hydroxyapatite.[159] In composition and structure, this layer strongly resembles the so-called hydroxyapatite in bone and it is therefore anticipated that osteoblasts should be capable of proliferating on this surface, and laying down fully formed bone, rather than fibrous capsule. This has been confirmed by experiments using rats as the animal model. Implanted A/W ceramic were found to bond so strongly through the interfacial apatite layer that they could not be separated.[160] Similar layers of bone-like apatite have been observed on Bioglass® and sintered hydroxyapatite, but they form at slower rates than the layer on he A/W glass ceramic, a fact which has been suggested as explaining the higher bioactivity of the latter materials.[161]

Under normal conditions, human body fluid is supersaturated with respect to apatite,[162] so that when apatite nuclei form, crystals can grow spontaneously. The chemical species that are capable of acting as nuclei are calcium and silicate ions while, by contrast, phosphate ions do not act as nuclei for apatite crystal growth. Hence, ceramics that release the former ions are capable of developing a surface apatite layer when exposed to human body fluid. This explains why glasses that do not contain P_2O_5 are able to develop this surface layer, and show greater bioactivity than glasses containing P_2O_5, even where the latter have compositions approximating to that of hydroxyapatite.[163]

This excellent bioactivity and biocompatibility to bone suggests that these materials are capable of a wide range of biomedical applications. Experimental studies have been carried out using rabbits or sheep as animal models, and these have shown that A/W ceramic is suitable for load-bearing applications in appropriate circumstances. Long-bones of rabbit[164] and vertebrae of sheep were augmented with A/W ceramic, both with acceptable results. A few clinical studies have been reported in human patients. For example, in one case, a female patient with breast cancer metastasis in the 10th thoracic vertebra had the affected vertebra removed completely and the adjacent ones partly excised, and replaced with A/W glass-ceramic prostheses. Results were excellent, and the patient survived a further 8 years with no recurrence of the bone cancer, nor other adverse effects.[165] To date, such applications have been limited, due to the relatively brittle nature of the material, and the restrictions this places on the extent to which it can be used to replace or augment load-bearing bones. Nonetheless, they show what this material is capable of achieving in the clinic, and suggest the direction in which future developments may lie.

8 Alumina

Alumina has been used in both the fabrication of femoral heads for artificial hip joints and as the root structures of implantable artificial teeth. In both applications it shows almost complete inertness, and has therefore been accepted reasonably readily for application within the body.

Alumina femoral heads are employed as an alternative to all-metal femoral components for a number of reasons. They are dimensionally stable, and give rise to little or no adverse tissue reactions, either local or systemic.[166] They also have the important advantage of reduced wear of UHMWPE acetabular caps. Alumina surfaces can be fabricated with high roundness and very low surface roughness, and these make it possible to achieve *in vitro* wear rates against UHMWPE of the order of 0.02 μm per 10^6 cycles, compared with metal/UHMWPE wear rates of between 0.02 and 0.07 μm per 10^6 cycles.[167] The main drawback of alumina is that it is brittle, like all ceramics, but unlike certain ceramics it cannot be toughened readily. The main approach to ensuring a resorbable strength of alumina femoral heads is to control the size and distribution of defects through careful design and production procedures.[168]

In practical terms, alumina ball heads are fitted to metal femoral stems with a tapered joint. The robustness of the fixation of the head on the stem is a function of the depth of penetration of the metal into the tapered mouth on the alumina ball. By carefully matching properties such as surface roughness, roundness and linearity, the fixation of the head to the stem can be strong, and the overall artificial joint safe for the patient. Since their introduction in the mid-1970s,[169] some two million alumina ball heads have been implanted worldwide, and results are very good. A number of studies have reported no failures, and several have reported failure rates below 1%.

The chemical composition of alumina femoral heads is typically at least 99.5% Al_2O_3, a composition that is specified in the relevant ISO standard.[170] Allowed impurities are silica and alkali metal and alkaline earth oxides.

Following retrieval, though, the composition may be different, one study reporting that the total oxide impurity in a recovered alumina ball head was 5.66%, though it was not clear whther this due to changes brought about by exposure to body fluids, or because the implanted material had been of inferior quality. Grain size of the alumina crystallites is also critical to the success of these components. For biomedical application, optimal properties are achieved by having a narrow distribution and small grain size,[171] typical sizes ranging from 0.2 to 5 μm. These features are achieved by careful control of the sintering process by which alumina powder is consolidated into a single bulk material.

Alumina has been used in dentistry as single tooth replacement materials[172] in which use they have a success rate at 5 years of 90%. Studies in monkeys have shown them to have excellent biocompatibility and to integrate well with the gingival tissue,[173] though comparable results have been difficult to demonstrate unambiguously in man.[174] In fact various types of tissue reaction have been found for alumina implants. For example, some investigators have found connective tissue,[175] while others have found mineralised bone.[176] These differences appear to arise from variations in the amount of motion between the implant and surrounding tissues, which in turn may be a consequence of variations in the time of occlusal loading after implantation.[177]

A detailed case study of an alumina implant has been published.[177] In this study, the implant was a replacement for a left lateral incisor, and was left *in situ* for 30 months. After this, microscopic examination revealed that the implant was covered with highly mineralised compact lamellar bone, and that the interface was free of connective tissue or inflammatory cells. In addition, osteocytes were observed very close to the bone–implant surfaces. These features were all indicative of the excellent biocompatibility of this alumina implant in contact with bone, and demonstrated that the implant did not interfere with the process of bone remodelling under clinical conditions.

9 Zirconia

Zirconia, ZrO_2, has been studied as an alternative to alumina in the fabrication of ceramic femoral heads for artificial hip joints.[178] Like alumina, it presents a lower friction bearing surface to the UHMWPE acetabular cap than any of the possible metal femoral heads, with the result that there is reduced loss of polymer through wear in clinical service.

Zirconia needs to have 3 mol% yttria incorporated into it in order to facilitate a process known as transformation toughening. What this means is that, when zirconia is heated in the region 1000–1100 °C, it undergoes a phase transformation from monoclinic crystal symmetry to tetragonal. The tetragonal phase is denser and tougher than the monoclinic phase, both of which are attractive features. However, they disappear on cooling to room temperature, when the material reverts to the monoclinic phase. To prevent this reversion to monoclinic crystal morphology, yttria is added in low concentrations and this stabilises the tetragonal phase, so that it remains the predominant phase when the zirconia is cooled to room temperature. To date, clinical results with yttria-stabilised zirconia have been generally good, though concerns have been expressed both about the long-term *in vivo* stability of the tetragonal phase with respect to reversion to monoclinic geometry and the inherent radioactivity of the material.[179]

The issue of *in vivo* stability has been contentious, and there have been conflicting *in vitro* results in the literature about the stability of ytrria-stabilised zirconia in solutions designed to mimic physiological conditions. Following storage in water or simulated body fluid, there was found to be very little difference between the samples. Both storage solutions were associated with the development of very small amounts (<1%) of monoclinic zirconia at the surface but this was not associated with any reduction in mechanical properties. On the other hand, ageing for long periods in Ringer's solution was found to promote the formation of considerably greater amounts of the monoclinic phase, and this resulted in severe reductions in strength. The latter work led to the conclusion that this material should not be used for implants.[180] In the end, the real issue is whether simulated body fluid or Ringer's solution most closely models the environment to which yttria-stabilised zirconia is exposed in the body, and the balance of opinion seems to favour the findings that this material is sufficiently stable for clinical use, despite the probably very minor surface conversion of zirconia from tetragonal to monoclinic. Certainly, the material has been used as the femoral head of hip implants for several years with success.

The properties of a commercial yttria-stabilised zirconia are shown in Table 3.12, together with properties a medical grade alumina, also fabricated for use as ceramic femoral heads.

The mechanical properties of the yttria-stabilised zirconia can be seen to be superior to those of the alumina, and these arise because of the toughening behaviour of the yttria. The mechanism by which ytrria toughens the zirconia is not fully clear, but it is thought to be through the transformation of the metastable tetragonal structure into the stable monoclinic structure in regions of high stress, for example the crack tip. This, being an energy absorbing process and resulting in a crack tip of altered crystal structure, then inhibits further growth of the crack.

The combination of superior physical properties and apparent *in vivo* stability of yttria-stabilised zirconia makes it an attractive material for medical

Table 3.12 *Properties of 3 mol% yttria-stabilised zirconia compared with those of alumina*

Property	Yttria-stabilised zirconia	Alumina
Bulk density (g cm^{-3})	6.06	3.95
Average grain size (μm)	0.35	1.53
Mean strength (MPa)	810	396
Elastic modulus (GPa)	220	394
Monoclinic phase	<1%	Not applicable

use. Its improved properties give it the design flexibility to be employed in a wider range of femoral head shapes and sizes than alumina and, as more clinical data become available on its performance, its applications may spread to other surgical applications. These are predicted to include knee joints, finger joints and spinal implants.[179]

10 Pyrolytic Carbon

Carbon is capable of existing in several crystalline forms, some of which possess excellent thrombo-resistance properties. Because of this, carbon is the preferred material to interface with flowing blood, and is used in a variety of applications, such as artificial heart valves and pacemaker electrodes.[181] Three types of carbon are used for this application, namely (i) low temperature isotropic pyrolytic carbon, LTI, (ii) vitreous carbon and (iii) ultra-low temperature isotropic carbon, ULTI. These all have disordered lattice structures, and are referred to collectively as turbostratic carbons. Although the structures are disordered, they are actually quite similar to the graphite structure. In graphite, layers of carbon atoms are covalently bonded to form two-dimensional hexagonal arrays. These layers in turn are relatively weakly held together by van der Waals forces.

A typical application is the use of silicon-alloyed LTI carbon as a coating or as a machined monolithic material from which artificial heart valves are made. This material not only possesses excellent haemocompatibility, it also is extremely durable, strong and resistant to wear.

LT1 pyrolytic carbon components are made by depositing carbon and silicon carbide onto a polycrystalline graphite substrate using chemical vapour deposition. This process employs a fluidised bed, as illustrated in Figure 3.7, with a mixed carrier gas consisting of a silane, helium and a hydrocarbon.

At the processing temperatures (1000–1500 °C), the hydrocarbons undergo cracking or pyrolysis, forming solid products which deposit on the substrate:

$$CH_3CH_2CH_3 \rightarrow 3C + 4H_2$$
$$CH_3SiCl_3 \rightarrow SiC + 3HCl$$

The silicon content of the final coating controls the fracture strength, wear, hardness and elastic modulus. It exists as the carbide, rather than in elemental form as a mixture with the carbon, and has been shown to have cubic crystal symmetry.

Vitreous carbons, on the other hand, can be prepared by the slow heating of polymers such as cellulosics or phenol–formaldehyde resins. Such heating initially yields volatile gases, which are allowed to escape, and the need to allow this limits the thickness of vitreous carbon coatings that can be obtained to about 7 mm. Further heating leads to the formation of the vitreous carbon as a solid phase material. Vitreous carbons contain a random array of crystallites about 5 nm in size and, though the denisty is low, this is not due to porosity, and the permeability of these materials is low.

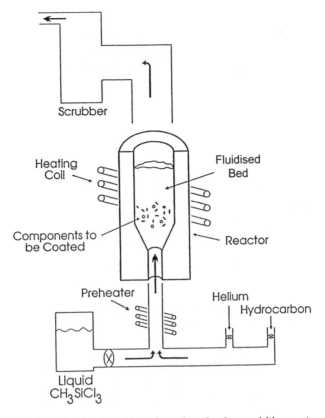

Figure 3.7 *Fluidised bed for the deposition of graphite for diamond-like coatings*

The most important feature of these materials from the biomedical point of view is their remarkably good haemocompatibility. Blood is a very difficult body fluid with which to interface a biomaterial, since it so readily coagulates, and may form potentially life-threatening thromboses. LT1 carbon, though, has excellent haemocompatibility, a feature that has not been fully explained. One possibility is that the smooth surface of the LT1 carbon allows selective adsorption of a passivating protein. Another is that the smoothness makes the surface inert to proteins in general, including those responsible for the process of clotting of the blood. They may thus be unable to deposit onto the surface. Either way, it is known that smooth or polished carbon surfaces are superior to rougher ones, since the latter permit cells to attach and act as the nuclei for thrombus development.

LT1 carbons are used to coat a variety of prosthetic heart valves. Such valves face difficult operating conditions, and must withstand cyclic loading, flexing and bending, surface wear and exposure to a potentially corrosive fluid for considerable periods of time. This is one of the most demanding applications for any biomaterial, and LT1 carbons have been shown over several years to be equal to these demands. It has been estimated that at least 200 000

LT1-carbon coated heart valve devices are currently in use, and though there seem to be no problems as far as biocompatibility is concerned, there have been questions about the resistance of these materials to long-term loading, and some reports of cyclic fatigue.[182] Further investigation is continuing on this latter subject.

11 References

1   R. Cotterill, *The Cambridge Guide to the Material World*, Cambridge University Press, UK, Cambridge, 1985.
2   D. F. Williams, *Concise Encyclopedia of Medical and Dental Materials*, xvii, Pergamon Press, UK, Oxford, 1990.
3   L. L. Hench and J. Wilson, *Introduction to Bioceramics*, World Scientific Publishers, Singapore, 1993.
4   I. D. Thompson and L. L. Hench, *Proc. Inst. Mech. Eng., Part H*, 1998, **212**, 127.
5   L. L. Hench, *Bioceramics*, 1994, **7**, 3.
6   R. Li, A. E. Clark and L. L. Hench, *J. Appl. Biomater.*, 1991, **2**, 231.
7   G. Vekinsis, M. F. Ashby and P. W. R. Beaumont, *J. Mater. Sci.*, 1993, **28**, 3221.
8   F. C. M. Dreissens, E. A. P. de Maeyer, E. Fernandez, M. G. Boltong, G. Berger, R. M. H. Verbeeck, M. P. Ginebra and J. A. Planell, *Bioceramics*, 1996, **9**, 231.
9   J. Lemaitre, *Innovations Tech. Biol. Med.*, 1995, **16**, 109.
10 W. E. Brown and L. C. Chow, *J. Dent. Res.*, 1983, **62**, 672.
11 E. Fernandez, F. J. Gil, M. P. Ginebra, F. C. M. Driessens, J. A. Planell and S. M. Best, *J. Mater. Sci.; Mater. Med.*, 1999, **10**, 177.
12 F. C. M. Driessens and R. M. H. Verbeeck, *Biominerals*, CRC Press, Boca Raton, FL, 1990.
13 E. A. P. de Maeyer, R. M. H. Verbeeck and D. E. Nassens, *Inorg. Chem*, 1993, **32**, 5709.
14 F. C. M. Driessens, J. A. Planell and F. J. Gil, in D. L. Wise, D. J. Trantolo, D. E. Altobelli, M. J. Yaszemski, J. D. Gresser and E. R. Schwartz (eds.), *Encylcopedic Handbook of Biominerals and Bioengineering, Part B, Applications*, Marcel Dekker, New York, 1995, pp. 855–877.
15 J. Lemaitre, A. Mirtchi and A. Mortier, *Silicates Ind. Ceram. Sci. Technol.*, 1987, **52**, 141.
16 L. C. Chow, *J. Ceram. Soc. Jpn. Int. Ed.*, 1992, **99**, 927.
17 M. T. Fulmer and P. W. Brown, *J. Mater. Sci.; Mater. Med.*, 1992, **3**, 299.
18 M. T. Fulmer and P. W. Browm, *J. Mater. Res.*, 1993, **8**, 1687.
19 P. W. Brown, J. Gulick and J. Dumm, *J. Biomed. Mater. Res.*, 1994, **28**, 27.
20 K. S. Tenhuisen and P. W. Brown, *J. Mater. Sci.; Mater. Med.*, 1994, **5**, 291.
21 A. A. Mirtchi, J. Lemaitre and E. Munting, *Biomaterials*, 1990, **11**, 83.
22 A. A. Mirtchi, J. Lemaitre and E. Munting, *Biomaterials*, 1991, **12**, 505.
23 J. Lemaitre, E. Munting and A. A. Mirtchi, *Rev. Stomatol. Chir. Maxillofac.*, 1992, **93**, 163.
24 B. R. Constanz, I. C. Ison, M. T. Fulmer, R. D. Poser, S. T. Smith, M. Van Wagoner, J. Ross, S. A. Goldstein, J. B. Jupiter and D. I. Rosenthal, *Science*, 1995, **267**, 1796.
25 E. Fernadez, J. A. Planell, S. M. Best and W. Bonfield, *J. Mater. Sci.; Mater. Med.*, 1998, **9**, 789.

26   E. A. P. De Maeyer, R. M. H. Verbeeck and C. W. J. Vercruysse, *J. Biomed. Mater. Res.*, 2000, **52**, 95.

27   W. E. Brown, L. W. Schroeder and J. S. Ferris, *J. Phys.Chem.*, 1979, **83**, 1385.

28   B. B. Tomazic, I. Meyer and W. E. Brown, *J. Cryst. Growth*, 1991, **108**, 670.

29   H. Monma and T. Kamiya, *J. Mater. Sci.*, 1987, **22**, 4247.

30   H. Monma and T. Kamiya, *J. Mater. Sci.*, 1980, **15**, 2428.

31   H. Monma, A. Makishima, M. Mitomo and T. Ikegami, *Nippon Seramikkusu Kyokai Gadujutsu Ronbushi*, 1988, **96**, 878.

32   O. Bermudez, M. G. Boltong, F. C. M. Driessens and J. A. Planell, *J. Mater. Sci.; Mater. Med.*, 1994, 5, 67.

33   O. Bermudez, M. G. Boltong, F. C. M. Driessens and J. A. Planell, *J. Mater. Sci.; Mater. Med.*, 1994, **5**, 160.

34   O. Bermudez, M. G. Boltong, F. C. M. Driessens and J. A. Planell, *J. Mater. Sci.; Mater. Med.*, 1994, **5**, 164.

35   Y. Fukase, E. D. Eanes, S. Takagi, L. C. Chow and W. E. Brown, *J. Dent. Res.*, 1990, **69**, 1852.

36   M. P. Ginebra, E. Fernandez, M. G. Boltong, O. Bermudez, J. A. Planell and F. C. M. Driessens, *Clin. Mater.*, 1994, **17**, 99.

37   E. Fernandez, M. P. Ginebra, O. Bermudez, M. G. Boltong and F. C. M. Driessens, *J. Mater. Sci. Lett.*, 1995, **14**, 4.

38   F. C. M. Driessens, J. A. Planell, M. G. Boltong, I. Khairoun and M. P. Ginebra, *Proc. Inst. Mech. Eng.*, Part H, 1998, **212**, 427.

39   O. Bermudez, M. G. Boltong, F. C. M. Driessens and J. A. Planell, *J. Mater. Sci.; Mater. Med.*, 1993, **4**, 389.

40   F. C. M. Driessens and R. M. H. Verbeeck, in C. de Putter, G. L. de Lange, K. de Groot and A. J. C. Lee (eds.), *Implant Materials in Biofunction*, Elsevier, Amsterdam, 1988, pp. 105–111.

41   F. C. M. Driessens, *Ann. NY Acad. Sci.*, 1988, **523**, 131.

42   F. C. M. Driessens, *Euroceramics*, 1989, **3**, 348.

43   M. M. A. Ramselaar, F. C. M. Driessens, W. Kalk, J. R. de Wyn and P. J. van Mullem, *J. Mater. Sci.; Mater. Med.*, 1991, **2**, 63.

44   J. Cavalheiro, R. Branco and M. Vasconselos, *Bioceramics*, 1991, **4**, 205.

45   M. Bohner, J. Lemaitre, K. Ohura and P. Hardouin, *Ceramic Trans.*, 1995, **48**, 245.

46   K. Ohura, M. Bohner, P. Hardouin, J. Lemaitre, G. Pasquier and B. Flautre, *J. Biomed. Mater. Res.*, 1996, **30**, 193.

47   M. R. Christoffersen, J. Christoffersen and W. Kibakzyc, *J. Cryst. Growth*, 1990, **106**, 349.

48   F. C. M. Driessens, M. G. Boltong, M. I. Zapatero, R. M. H. Verbeeck, W. Bonfield, O. Bermudez, E. Ferandez, M. P. Ginebra and J. A. Planell, *J. Mater. Sci.; Mater. Med.*, 1995, **6**, 272.

49   F. C. M. Driessens, M. G. Boltong, J. A. Planell, O. Bermudez, M. P. Ginebra and E. Fernandez, *Bioceramics*, 1993, **6**, 469.

50   F. C. M. Driessens, M. G. Boltong, I. Khairoum, M. P. Ginebra, E. Fernandez and J. A. Planell, *Bioceramics*, 1997, **10**, 279.

51   C. D. Friedman, P. D. Constantinos, K. Jones, L. C. Chow, H. J. Pelzer and G. A. Sisson, *Arch. Otolar. Head Neck Surg.*, 1991, **117**, 385.

52   Y. Unezaki, H. Rynmon, H. Inoue, H. Okuda, H. OOnishi, K. Minamigana and F. Sugihara, *Bioceramics*, 1996, **9**, 251.

53   C. A. Beevers and D. B. McIntyre, *Miner. Mag.*, 1956, **27**, 254.

54 R. A. Young and J. C. Elliot, *Arch. Oral Biol.*, 1966, **11**, 699.
55 E. C. Moreno, M. Kresak and R. T. Zahradnik, *Caries Res.*, 1977, **11**, 142.
56 Y. Doi, Y. Morwaki, T. Aoba, M. Okazaki, J. Takahashi and K. Joshin, *J. Dent. Res.*, 1982, **61**, 429.
57 W. E. Brown and L. Chow, *Ann. Res. Mater. Sci.*, 1976, **6**, 213.
58 M. J. Glimcher, *Phil. Trans. R. Soc. London*, 1984, **B304**, 479.
59 R. Z. LeGeros and J. P. LeGeros, in L. L. Hench and J. Wilson, *An Introduction to Bioceramics*, World Scientific, Singapore, 1993, Chapter 9.
60 R. Z. LeGeros, O. R. Trautz, J. P. LeGeros and W. P. Shira, *Science*, 1967, **155**, 1409.
61 R. Z. LeGeros, *Prog. Crystal Growth Charact.*, 1981, **4**, 1.
62 M. Yoshikowa, T. Toda, H. Oonishi, S. Kushitani, E. Yasukawa, S. Hayani, Y. Mandai and F. Sugihara, *Bioceramics*, 1994, **6**, 187.
63 A. C. Lawson and J. T. Czernuska, *Proc. Inst. Mech. Eng., Part H*, 1998, **212**, 413.
64 K. E. Watson, K. S. Tenhuisen and P. W. Brown, *J. Mater. Sci.; Mater. Med.*, 1999, **10**, 205.
65 Y. Matsuya, J. M. Antonucci, S. Matsuya, S. Takagi and L. C. Chow, *Dent. Mater.*, 1996, **12**, 2.
66 Y. Doi, *Cells Mater.*, 1997, **7**, 111.
67 C. Knabe, F. C. M. Driessens, J. A. Planell, R. Gildenhaar, G. Berger, D. Reif, R. Fitzner, R. J. Radlanski and U. Gross, *J. Biomed. Mater. Res.*, 2000, **52**, 498.
68 K. de Groot, R. G. T. Geesink, C. P. A. T. Klein and P. Serekin, *J. Biomed. Mater. Res.*, 1987, **21**, 1375.
69 R. G. T. Geesink, K. de Groot and C. P. A. T. Klein, *Clin. Orthop. Rel. Res.*, 1987, **225**, 147.
70 K. de Groot, J. G. C. Wolke and J. A. Jansen, *Proc. Inst. Mech. Eng., Part H*, 1998, **212**, 137.
71 S. D. Cook, K. A. Thomas, J. F. Kay and M. Jarcho, *Clin. Orthop. Rel. Res.*, 1988, **232**, 225.
72 J. G. C. Wolke, J. M. A. de Blieck-Hogervorst, W. J. A. Dhert, C. P. A. T. Klein and K. de Groot, *J. Thermal Spray Technol.*, 1992, **1**, 75.
73 J. Disam, K. Luebbers, K. Neudert and A. Sickinger, *J. Thermal Spray Technol.*, 1994, **3**, 142.
74 H. D. Steffens and M. Brune, *J. Thermal Spray Technol.*, 1995, **4**, 85.
75 J. G. C. Wolke, J. M. A. de Blieck-Hogervorst, T. G. Kraak, W. Herlaar and K. de Groot, *Proc. 4th Nat. Thermal Spray Conf.*, Pittsburgh, PA, 1991, p. 481.
76 H. Oguchi, K. Ishikawa, S. Ojima, Y. Hirayama, K. Seto and G. Eguchi, *Biomaterials*, 1992, **13**, 471.
77 P. Ducheyne, W. van Raemdonck, J. C. Heughebaert and M. Heughebaert, *Biomaterials*, 1986, **7**, 97.
78 P. Ducheyne, S. Radin, M. Heughebaert and J. C. Heughebaert, *Biomaterials*, 1990, **11**, 244.
79 S. Ban, S. Maruno, N. Arimoto, A. Harada, M. Matsuura and J. Hasegawa, *Bioceramics*, 1994, **7**, 261.
80 P. Ducheyne, L. L. Hench, A. Kagan II, M. Martens, A. Bursens and J. C. Mulier, *J. Biomed. Mater. Res.*, 1980, **14**, 225.
81 T. Hanawa and M. Ota, *Appl. Surf. Sci.*, 1992, **55**, 269.
82 E. Leitao, M. A. Barbosa and K. de Groot, *J. Mater. Sci.; Mater. Med.*, 1995, **6**, 849.

83 D. R. Cooley, A. F. van Dellen, J. O. Burgess and S. Windeler, *Prosthet. Dent.*, 1992, **67**, 93.

84 J. A. Jansen, J. G. C. Wolke, S. Swann, J. P. C. M. van der Waerden and K. de Groot, *Clin. Oral Impl. Res.*, 1993, **4**, 28.

85 J. G. C. Wolke, K. van Dijk, H. G. Schaeken, K. de Groot and J. A. Jansen, *J. Biomed. Mater. Res.*, 1994, **28**, 1477.

86 H. Solnick-Legg and K. Legg, *MRS Bull.*, 1989, 27.

87 J. R. Stevenson, H. Solnick-Legg and K. O. Legg, *Mater. Res. Soc., Symp. Proc.*, 1989, **110**, 715.

88 C. M. Cotell, *Appl. Surf. Sci.*, 1993, **69**, 140.

89 R. K. Singh, F. Qian, V. Nagabushnam, R. Damodaran and B. M. Moudgil, *Biomaterials*, 1994, **15**, 522.

90 K. A. Thomas, C. D. Cook, R. J. Ray and M. Jarcho, *J. Arthroplasty*, 1989, **4**, 43.

91 J. A. Jansen, J. P. C. M. van der Waerden, J. G. C. Wolke and K. de Groot, *J. Biomed. Mater. Res.*, 1991, **25**, 973.

92 W. J. A. Dhert, C. P. A. T. Klein, J. G. C. Wolke, E. A. van der Velde, K. de Groot and P. M. Rozing, *J. Biomed. Mater. Res.*, 1992, **25**, 1183.

93 W. J. A. Dhert, C. P. A. T. Klein, J. A. Jansen, E. A. van der Velde, R. C. Vriesde, K. de Groot and P. M. Rozing, *J. Biomed. Mater. Res.*, 1993, **27**, 127.

94 K. Soballe, E. S. Hansen, H. B. Brockstedt-Rasmussen and C. Bunger, *J. Bone Jt. Surg.*, 1993, **75B**, 270.

95 R. G. T. Geesink, *Clin. Orthop. Rel. Res.*, 1990, **261**, 39.

96 L. Heling, R. Heindel and B. Merin, *J. Oral Implantol.*, 1981, **9**, 548.

97 C. P. A. T. Klein, P. Patka, H. B. M. van der Lubbe, J. G. C. Wolke and K. de Groot, *J. Biomed. Mater. Res.*, 1991, **25**, 53.

98 F. C. M. Dreissens, *Proc. NY Acad. Sci.*, 1988, **523**, 131.

99 S. L. Rowles, *Bull. Soc. Fr.*, 1968, 1797.

100 J. D. Bruijn, Y. P. Bovell, C. P. A. T. Klein, K. de Groot and C. A. van Blitterswijk, *Biomaterials*, 1994, **15**, 543.

101 J. F. Osborne, *Biomed. Technol.*, 1987, **32**, 177.

102 E. Munting, in *Hydroxyapatite Coatings in Orthopaedic Surgery*, ed. R. G. T. Geesink and M. T. Manley, Raven Press, New York, 1993, pp. 13–20.

103 M. J. Filiaggi, N. A. Coombs and R. M. Pilliar, *J. Biomed. Mater. Res.*, 1991, **25**, 1211.

104 F. J. Kummer and W. L. Jaffe, *J. Appl. Mater.*, 1992, **3**, 211.

105 R. L. Reis, F. J. Monteiro and G. W. Hastings, *J. Mater. Sci.; Mater. Med.*, 1994, **5**, 457.

106 R. D. Bloebaum and J. A. Dupont, *J. Arthroplasty*, 1993, **8**, 195.

107 R. D. Bloebaum, D. Beeks, L. D. Dorr, C. G. Savory, J. A. Dupont and A. A. Hofman, *Clin. Orthop.*, 1994, **298**, 19.

108 K. de Groot, C. P. A. T. Klein and A. A. Driessen, *J. Head Neck Pathol.*, 1985, **4**, 90.

109 H. W. Denissen, K. de Groot, A. A. Driessen, J. G. C. Wolke, J. G. J. Peelen, H. J. A. van Dijk, A. P. Gehring and P. J. Klopper, *Sci. Ceram.*, 1980, **10**, 63.

110 S. Best, B. Sim, M. Kayser and S. Downes, *J. Mater. Sci.; Mater. Med.*, 1997, **8**, 97.

111 J. J. Klawitter and S. F. Hulbert, *J. Biomed. Mater. Res.*, 1972, **6**, 347.

112 R. E. Holmes, V. Mooney, R. Bucholz and A. Tencer, *Clin. Orthop. Rel. Res.*, 1984, **188**, 252.

113 P. Li, C. Ohtsuki, T. Kokubo, K. Nakanishi, N. Soga and K. de Groot, *J. Biomed. Mater. Res.*, 1994, **28**, 7.

114 A. S. Posner, *Physiol. Rev.*, 1969, **49**, 760.

115 I. Abraham and J. C. Knowles, *J. Mater. Chem.*, 1994, **4**, 185.

116 N. C. Blumenthal and A. S. Posner, *Calc. Tiss. Int.*, 1984, **36**, 439.

117 K. H. Lau, A. Yoo and S. Ping-Wang, *Mol. Cell Biochem.*, 1991, **105**, 93.

118 J. H. Kuhne, R. Bartl, B. Frish, C. Hanmer, V. Jansson and M. Zimmer, *Acta Orthop. Scand.*, 1994, **65**, 246.

119 P. S. Eggli, W. Muller and R. K. Schenk, *Clin. Orthop. Rel. Res.*, 1988, **232**, 127.

120 W. J. A. Dhert, C. P. A. T. Klein, J. G. C. Wolke, E. A. van der Velde, K. de Groot and P. M. Rozing, *J. Biomed. Mater. Res.*, 1991, **25**, 1183.

121 K. A. Hing, S. M. Best and W. Bonfield, *J. Mater. Sci.; Mater. Med.*, 1999, **10**, 135.

122 J. D. Currey, *Clin. Orthop. Rel. Res.*, 1970, **73**, 210.

123 K. A. Hing, S. M. Best, K. E. Tanner, W. Bonfield and P. A. Revell, *J. Mater. Sci.; Mater. Med.*, 1999, **10**, 663.

124 H. Denissen, C. Mangano and G. Cenini, *Hydroxylapatite Implants*, Piccin Nuova Libraria SPA, India, 1985.

125 E. Hayek and H. Newesely, *Inorg. Synth.*, 1963, **7**, 63.

126 R. Z. LeGeros, *Calcium Phosphates in Oral Biology and Medicine, Monographs in Oral Sciences*, ed. H. Myers, Vol. 15, S, Karger, Basle, Switzerland, 1991.

127 J. H. Quinn and J. N. Kent, *Oral Surg.*, 1984, **58**, 511.

128 J. Wilson and G. E. Merwin, *J. Biomed. Mater. Res., Appl. Biomater.*, 1988, **22**, 157.

129 P. N. Calgut, I. M. Waite and S. M. B. Tinkler, *Clin. Mater.*, 1990, **6**, 105.

130 J. Huang, L. Di Silvio, M. Wang, K. E. Tanner and W. Bonfield, *J. Mater. Sci.; Mater. Med.*, 1997, **8**, 775.

131 P. Ducheyne, C. S. Kim and S. R. Pollack, *J. Biomed. Mater. Res.*, 1992, **26**, 147.

132 M. Gorbunoff, *J. Anal. Biochem.*, 1984, **136**, 425.

133 S. Niwa, K. Sawai, S. Takahashi, H. Tagai, M. Ono and Y. Fukuda, *Biomaterials*, 1980, **1**, 65.

134 M. Gregoire, I. Orly and J. Menanteau, *J. Biomed. Mater. Res.*, 1990, **24**, 163.

135 F. B. Bagambisa, U. Joos and W. Schilli, *Int. J. Oral Maxillofac. Implants*, 1990, **5**, 217.

136 E. Nery, R. Z. LeGeros and K. Lynch, *J. Periodontol.*, 1992, **63**, 729.

137 L. L. Hench and O. Andersson, in *Introduction to Bioceramics, Advanced Series in Ceramics*, Vol. 1, ed. L. L. Hench and J. Wilson, World Scientific, Singapore, 1993, Chapter 3.

138 L. L. Hench, *J. Am. Ceram. Soc.*, 1991, **74**, 1487.

139 L. L. Hench and D. E. Clark, *J. Non-Cryst. Solids*, 1978, **28**, 83.

140 O. H. Andersson, *J. Mater. Sci.; Mater. Med.*, 1992, **3**, 326.

141 N. C. Blumenthal and A. S. Posner, *Calcif. Tiss. Int.*, 1984, **36**, 439.

142 L. D. Quarles, *Miner. Electrolyte Metab.*, 1991, **17**, 233.

143 C. Oktsuki, T. Kokubo, K. Takatsuka and T. Yamamuno, *Nippon Seram. Kyokai Gak. Ron.*, 1991, **91**, 1.

144 O. H. Andersson and I. Kangasniemi, *J. Biomed. Mater. Res.*, 1991, **25**, 1019.

145 A. E. Clark, C. G. Pantano and L. L. Hench, *J. Am. Ceram. Soc.*, 1976, **58**, 37.

146 C. Y. Kim, A. E. Clark and L. L. Hench, *J. Non-Cryst. Solids*, 1989, **113**, 195.

147 L. L. Hench, *J. Am. Ceram. Soc.*, 1991, **74**, 1487.

148 J. Wilson and D. Noletti, in P. Christel, A, Meunier and A. J. C. Lee, *Biological*

and *Biomechanical Performance of Biomaterials*, Elsevier Science Publishers, Amsterdam, 1986, pp. 99–104.

149 J. Wilson and G. H. Pigott, *J. Biomed. Mater. Res.*, 1981, **15**, 805.

150 H. R. Stanley, M. B. Hall, F. Colaizzi and A. E. Clark, *J. Prosthet. Dent.*, 1987, **58**, 607.

151 J. Wilson and S. B. Low, *J. Appl. Biomater.*, 1992, **3**, 123.

152 T. Kokubo, M. Shigematsu, Y. Nagashima, M. Tashiro, T. Yamamuro and S. Higashi, *Bull. Inst. Chem. Res., Kyoto Univ.*, 1982, **60**, 260.

153 T. Kokubo, S. Ito, S. Sakka and T. Yamamuro, *J. Mater. Sci.*, 1986, **21**, 536.

154 T. Kokubo, H. Kushitani, C. Ohtsuki, S. Sakka and T. Yamamura, *J. Mater. Sci.; Mater. Med.*, 1992, **3**, 79.

155 T. Kokubo, S. Ito, M. Shigematsu, S. Sakka and T. Yamamuro, *J. Mater. Sci.*, 1985, **20**, 2001.

156 K. Ono, T. Yamamuro, T. Nakamura and T. Kokubo, *Biomaterials*, 1990, **11**, 265.

157 T. Kitsugi, T. Nakamura, T. Yamamuro, T. Kokubo, T. Shibuya and M. Takagi, *J. Biomed. Mater. Res.*, 1987, **21**, 1255.

158 M. Neo, S. Kotani, Y. Fujita, T. Nakamura, T. Yamamuro, Y. Bando, C. Ohtsuki and T. Kokubo, *J. Biomed. Mater. Res.*, 1992, **26**, 255.

159 T. Kokubo, S. Ito, Z. Huang, T. Hayashi, S. Sakka, T. Kitsugi and T. Yamamura, *J. Biomed. Mater. Res.*, 1987, **24**, 331.

160 T. Kitsugi, T. Yamamuro, T. Nakamura, T. Kokubo, M. Takagi, T. Shibuya, H. Takeuchi and M. Ono, *J. Biomed. Mater. Res.*, 1987, **21**, 1109.

161 T. Kokubo, H. Kushitani, Y. Ebisawa, T. Kigushi, S. Kotani, K. Oura and T. Yamamuro, in H. Oonishi and K. Sawai, *Bioceramics*, Vol. 1, Ishiyaku EuroAmerica, Tokyo, 1989, pp. 157–162.

162 W. Neuman and M. Neuman, in *The Chemical Dynamics of Bone Mineral*, University of Chicago Press, Chicago, 1958, p. 158.

163 K. Ohura, T. Nakamura, T. Yamamuro, T. Kokubo, Y. Ebisawa, Y. Kotoura and M. Oka, *J. Biomed. Mater. Res.*, 1991, **25**, 357.

164 T. Kigushi, T. Yamamuro and T. Kokubo, *J. Bone Jt. Surg.*, 1988, **71A**, 264.

165 T. Yamamuro, T. Nakamura, S. Higashi, R. Kasai, Y. Kakutani, T. Kitsugi and T. Kokubo in K. Atsumi, M. Maekawa and K. Ota, *Progress in Artificial Organs– 1983*, IASO Press, Cleveland, OH, 1984, pp. 810–814.

166 C. Piconi, M. Labanti, G. Magnani, M. Caporale, G. Maccauro and G. Magliocchetti, *Biomaterials*, 1999, **20**, 1637.

167 I. C. Clarke and G. Willmann in *Bone Implant Interface*, ed. U. C. Cameron, Mosby Publishers, St Louis, MO, 1994, pp. 203–252.

168 E. Dorre and H. Hubner, *Alumina*, Springer, Berlin, 1984.

169 P. M. Boutin, *Rev. Chir. Orthop.*, 1975, **58**, 229.

170 International Organisation for Standardisation, Ceramic materials based on high purity alumina, ISO 6474, 1981.

171 G. Heimke in *Ceramics in Surgery*, ed. P. Vincenzini, Elsevier, Amsterdam, 1983, pp. 33–41.

172 T. Albrektsson and L. Sennerby, *J. Clin. Periodontol.*, 1991, **18**, 474.

173 L. Zetterqvist, G. Anneroth and A. Nordenram, *Int. J. Maxillofac. Implants*, 1991, **6**, 285.

174 G. Anneroth, A. R. Ericsson and L. Zetterqvist, *Swed. Dent. J.*, 1990, **14**, 63.

175 M. Takuma, S. Harada, F. Kurokawa, F. Takashima, S. Miyauchi and T. Maruyama, *J. Osaka Univ. Dent. Sch.*, 1987, **27**, 121.

176 D. E. Steflik, A. L. Sisk and G. R. Parr, *J. Biomed. Mater. Res.*, 1993, **27**, 791.

177 A. Piatelli, G. Podda and A. Scarano, *Biomaterials*, 1996, **17**, 711.
178 T. Tateishi and H. Yunoki, *Clin. Mater.*, 1993, **12**, 219.
179 J. W. Stuart, I. C. Alexander and A. W. Pryce, in *21st Century Ceramics*, ed. D. P. Thompson and H. Mandal, British Ceramic Proceedings No 55, Institute of Materials, London, 1996, pp. 113–121.
180 I. Thompson and R. D. Rawlings, *Biomaterials*, 1990, **11**, 505.
181 R. H. Dauskardt and R. O. Ritchie, in *An Introduction to Bioceramics*, ed. L. L. Hench and J. Wilson, World Scientific, Singapore, 1993, Chapter 14.
182 R. O. Ritchie, R. H. Dauskardt and F. J. Pennisis, *J. Biomed. Mater. Res.*, 1992, **26**, 69.

CHAPTER 4
Metals

1 Introduction

A number of metallic alloys have found extensive use as biomaterials. They include cobalt–chromium alloys, which usually have small amounts of molybdenum in them, stainless steel, and titanium and their alloys, all of which have been used in orthopaedics. In addition, the complex silver–tin–copper amalgams have been used for well over 150 years in dentistry as filling materials, and gold and other precious metal alloys are used to fabricate crowns. In addition, titanium and its alloys are used as dental implants.

For hip replacement surgery, the standard materials for the femoral stem have been one of the following: (i) cobalt–chromium with small amounts of molybdenum and possibly nickel, (ii) 316L stainless steel or (iii) titanium with 6% aluminium and 4% vanadium, generally written Ti-6Al-4V. Cobalt–chromium and 316L stainless steel were the first metals to be used for such applications,[1] and were for a long time considered to be well tolerated by the body, even though corrosion products were frequently found in tissues surrounding the implant, or even at sites somewhat removed from the location of the implant.[2,3] These days, however, there is an increase in use of titanium alloys, and new alloys are being developed in attempts to improve on the excellent biocompatibility of the well established alloys such as Ti-6Al-4V. Their biocompatibility in bone contact applications arises partly from their relative lack of corrosion and of the low cytotoxicity of any corrosion products released, and also because their moduli are much lower than those of other possible metals for this purpose. Using low modulus materials reduces the mismatch between the modulus of the implant and that of the bone, which in turn reduces the so-called stress shielding phenomenon, which can be responsible for potentially damaging resorption of bone on the inner surface of the natural femur.

An indication of the properties of the three possible metal systems used in orthopaedics are given in Table 4.1. Some of these alloys also find use outside orthopaedics. For example, cobalt–chromium alloys are used in dentistry, and 316L stainless steel is used in temporary devices, such as fracture plates and

Table 4.1 *Characteristics of metal femoral pin components (from Long and Rack[4])*

	Stainless steel	*Cobalt–chromium*	*Titanium alloys*
Composition (%)	Fe + Cr (17–20); Ni (12–14); Mo (2–4)	Co + Cr (19–30); Mo (0–10); Ni (0–37)	Ti + Al (6); V (4) + possibly Nb (7)
Elastic modulus	200 GPa	200–230 GPa	55–110 GPa
Advantages	Low cost Ease of processing	Low wear Low corrosion Resists fatigue	Biocompatible Low corrosion Low modulus Resists fatigue
Disadvantages	High modulus Poor long term properties	High modulus Low bio-compatibility	Low shear strength Expensive

screws, which are left *in situ* while damaged bone repairs itself, and then removed to allow healing to become complete.

The elastic modulus for all three of these alloys is almost independent of their processing history and consequent microstructure.[5] On the other hand, yield stress and ultimate tensile strength are both extremely sensitive to grain size, porosity and phases present. Generally, optimum properties are obtained where grain size is a minimum, and there is little or no porosity. In order to obtain uniform properties, grain size should not simply be minimal, it should also itself be uniform. The stainless steel is highly dependent on its processing history; it is ductile in the annealed condition, but changes markedly on being cold worked, with its yield stress and tensile strength increasing greatly. As a result, a wide range of yield stresses can be obtained, and though the values quoted vary, a range of from 207 to 1160 MPa has been reported.[6] The properties of cobalt–chromium can also be altered by cold working, though over a slightly narrower range, *i.e.* yield stresses being changed from 430 to 1028 MPa have been reported.[6] Both stainless steel and cobalt–chromium alloys are cold worked to make them suitable for use as implants; hence the yield stresses of orthopaedic devices lie towards the upper end of these ranges.

Titanium and its alloys are not sufficiently ductile in the annealed condition to allow them to be cold worked. However, even without being cold worked, these alloys have yield stresses in the range 780–1050 MPa, which compare with the upper end of the ranges for the other two classes of orthopaedic alloy. For all three types of alloy, the yield stress exceeds by some way the yield stress of cortical bone, so all three are more than satisfactory in terms of strength for use in joint replacement surgery.

An important consideration in the use of any of these metals is that the body is a corrosive environment. In the body, metals encounter extracellular fluid, an electrolyte solution that readily promotes electrochemical corrosion. This fluid may contain, in addition to electrolytes, biochemical species such as proteins or amino acids, and these can enhance corrosion. There have been a number of studies that have demonstrated that patients with metallic hip and knee prostheses develop high concentrations of metal ions in their body fluids, including in the serum and urine,[7-9] and these studies have raised concerns about the possibility that such metal ions may be responsible for a variety of adverse bone and tissue conditions that in turn lead to bone necrosis and failure of the implanted joint.[10,11]

Metallic corrosion in the body has two possible modes, one active, the other passive.[12] The term active corrosion implies that the body's defence mechanisms play an active role in promoting corrosion. This may be through altering the kinetics of the corrosion processes, or through the introduction of completely new corrosion reactions. By contrast, the passive mode of corrosion refers to those processes which occur through mechanisms that are the same as those in non-living environments. These corrosion processes are driven only by the physicochemical aspects of the metal and electrolyte medium, with no enhancement through active intervention of the body's biochemistry.

Corrosion not only reduces the quality and ultimately the strength of the metallic implant, it also introduces potentially toxic by-products in the form of metal ions released into the body fluids. These may affect the local environment, for example by promoting an adverse response such as inflammation or discoloration in the tissues adjacent to the implant. Alternatively, they can affect the body at sites remote from the implant.[1] Because of these adverse effects, much use is made of passivating treatments, including the use of alloys of noble metals, which remain unreacted under physiological conditions.

Gold and platinum are used, either as pure metals, or as the main component of alloys (*e.g.* Au–Ag, Au–Cu, Pt–Pd), for example in dentistry. The drawback of this approach is that noble metals are usually soft and weak, as well as expensive, and this limits their suitability for structural applications within the body. A less expensive alternative is to use passivated alloys of metals such as titanium or chromium. Alloys of this type include Ti–Al–V and Co–Cr, as well as a small number of stainless steels. Commercially pure titanium, cpTi, is also used, this being almost pure titanium with small amounts of oxygen (below 0.5%) in it.[13] In all of these cases, corrosion resistance within the body is reasonably good, especially with titanium and cobalt–chromium.

The chemistry, including the corrosion behaviour, of key alloys used as biomaterials will be considered throughout this chapter. However, before doing so, a few words are needed about the effect of the physiological environment. As has been mentioned, this is typically a source of species such as proteins, and it is known that the corrosion behaviour of certain metals is susceptible to being altered by the presence of proteins.[14] Those metals most affected are those whose ions are capable of bonding to proteins in solution, such as cobalt, silver and copper. Such metals can have their corrosion rates increased by an order of magnitude when proteins such as albumin are present. By contrast, metals such as titanium or chromium, which are covered by a strongly adherent oxide layer, do not seem to be affected to any significant extent by the presence of proteins.

2 Metal Ions in the Body

The effect of corrosion is that metallic ions are released into the body, and this may lead to local inflammatory response, or to more profound reactions, *i.e.* either mutagenesis or carcinogenesis. Although there is a link between these latter two biological effects, they are not synonymous, though most mutagens are carcinogens and *vice versa.*[15]

A mutagenic agent is one that brings about mutations, *i.e.* which causes the loss or appearance of a characteristic that can be inherited by subsequent generations. This requires that the agent interacts with the DNA of the organism and causes some sort of primary lesion in that DNA. Often, when such lesions occur, they can be repaired by the repair enzymes of the cell, but if there is some deficiency or overload of these enzymes, the lesion will persist. When the cells divide, this error will be replicated, and result in transformation into a mutation. The numbers of such lesions need to be small, or the cell will die, either because its vital functions are impaired, or because it will experience so-called apoptosis, or programmed cell death. On the other hand, cells with only small numbers of such lesions may be viable. They are therefore able to survive, replicate, and pass on the error on their DNA.

Mutations may affect either genes or chromosomes. Gene mutations are essentially of two types:

(i) Base pair substitutions. In this type of mutation, one base pair is replaced by another base pair, leading to a change in the genetic code.
(ii) Removal or addition of base pairs (so-called frameshift mutations), which results in changes to the encoded genetic information.

Chromosomal mutations are also of two types:

(i) Structure abnormalities, which arise due to a break in a strand of DNA, and which lead to loss of genetic information or to gene malfunctions.
(ii) Numerical abnormalities, where interaction of proteins leads to the formation of a reaction product with the DNA, and results in a change in the number of chromosomes. At the molecular level, these particular changes are mediated through the SH groups on the proteins, and these functional groups are particularly susceptible to interaction with any metal ions within the cells.

A variety of tests can be carried out to assess the mutagenicity/carconigenicity of metals. Of these, the most important are the chromosomal mutation tests. With this type of test, the effect of metal ions on the structure and number of the chromosomes can be determined reasonably readily. Studies are usually carried out using cultures of human lymphocyte cells, and these are exposed to the metal salt of interest, allowed to divide, and then treated with colcemide to arrest further cell division and subject the cells to hypotonic shock. In this state, they are spread on slides, stained and examined under the microscope. For a metal salt to be considered mutagenic, a significant number of cells should exhibit breaks in the chromosomes, and the effect should be dependent on dose.

As a result of studies of this type, there is now a considerable body of information on the proven or potential carcinogenicity of metals in humans. In a few cases, there is insufficient data to classify metals, but many of the metals used in biomedical alloys have been classified, and these will now be considered further.

Metals and their derivatives known to be carcinogenic in humans[16,17] are (i) nickel derivatives, (ii) hexavalent chromium, (iii) cadmium and its derivatives and (iv) beryllium and its derivatives. In all cases, epidemiological studies have demonstrated an increase in tumour incidence when these metals or their derivatives are present in the body. On the basis of proven mutagenicity only, cobalt and metallic nickel are considered potential carcinogens in humans.

Results for some metals are ambiguous, or lacking. For example, tin[18] and copper[19] in the 2+ oxidation state seem to be only slightly if at all mutagenic. Iron(II) is also almost completely safe, though it may bring about some genotoxic effects through the interference with endogenous hydrogen peroxide, and the resulting production of hydroxy free radicals:[20]

$$H_2O_2 + Fe^{2+} \leftrightarrow Fe^{3+} + OH^{\cdot} + OH^{-}$$

This reaction, though, does not result in either mutagenicity of carcinogenicity, and is readily brought under control within the body. Consequently, Fe^{2+} ions are not seen as presenting any risks when iron is used as the basis of biomedical devices.

Gold and platinum do not seem to have been studied for their mutagenicty/carcinogenicity, probably because they are so inert and strongly resist corrosion, even under the conditions prevailing within the body. Their uses are predominantly dental, where their relatively high price is offset by the very small volumes needed to fulfil their repair function. Silver, in the other hand, is known to corrode within the body,[21] and silver ions are well known to have strong inhibitory and bactericidal effects. It acts through the high affinity of the Ag^{+} ion for SH groups in proteins, and when this interaction occurs, the proteins become deactivated.[22] In cultures of bacteria this has the effect of causing the molecules of DNA to condense,[23] thus inhibiting DNA repair. However, these effects do not seem to induce genetic mutations in mammalian cells.[24,25]

Other metals, such as aluminium and vanadium, have known adverse effects within the body. These effects are not necessarily in terms of mutagenicity or carcinogenicity, but include both short-term and long-term damage. These metals are, however, relatively minor components of biomedical alloys, and their effects will be considered in more detail in the sections on specific alloys which follow.

As we have already seen, the main metals used in orthopaedics are 316L stainless steel, cobalt–chromium–molybdenum and titanium alloys and each of these is now discussed in detail. The chapter concludes with an account of the chemistry and uses of precious metals and their alloys, and of amalgam, both of which are mainly used in dentistry. The biological requirements for this application are generally less severe than for other surgical applications, since biomaterials used in repair or replacement rarely come into direct contact with either bone or blood. Instead, they are relatively isolated from the more sensitive tissues of the body. However saliva is corrosive, and consequently there are issues of biocompatibility and toxicity to consider in the use of materials within dentistry.

3 Cobalt–Chromium Alloys

There are a number of cobalt–chromium alloys that are used as biomaterials, and their applications include as orthopaedic devices (femoral pins and heads) and as dental prostheses.[26] Cobalt is usually the main component, and in addition to chromium there can be substantial amounts of molybdenum and/ or nickel. All cobalt–chromium alloys are strongly passivated, as a result of the existence of the tenacious chromium oxide coating. However, corrosion resistance does vary slightly between the different alloys, with nickel conferring particularly good corrosion resistance. Cobalt–chromium alloys are generally reasonably inert in the body, especially over short time intervals. Like titanium alloys, though, there is a small but finite corrosion current in physiological saline, resulting in losses of about 30 μg cm^{-2} of metals into the surrounding tissues. Bulk cobalt–chromium implants tend to become surrounded by fibrous tissue and to show no evidence of acute or chronic inflammation or foreign body reaction.[27]

In general, cobalt–chromium alloys are resistant to pitting and crevice corrosion, even within the body.[27] By contrast, relatively little is known about their susceptibility to stress corrosion cracking or corrosion fatigue. There have, though, been reports of fatigue failures *in vivo* of cobalt–chromium casting alloys for orthopaedics, an effect that may be assumed to arise from the presence of chloride ions in the surrounding body fluids under normal physiological conditions.

Cobalt–chromium alloys may undergo fretting corrosion quite readily.[27] The process of fretting is a mechanical one and involves rubbing in the form of a prolonged series of cyclic micro-movements. The result is localised damage to one or both surfaces. In fretting corrosion, the process continually exposes new surfaces, and these undergo oxidation. The fretting debris that becomes trapped between the surfaces enhances the rate of surface damage and exposure of new metal, and the whole process leads to loss of metal from the assembly. Detailed study of possible fretting corrosion under short-duration *in vitro* conditions failed to reveal any such damage,[25] which was taken as evidence that such corrosion would probably not occur under clinical conditions.

Table 4.2 *Mean concentrations* (mg cm^{-3}) *of cobalt and chromium in serum (from Granchi et al.*[34]*) [standard errors in brackets]*

	Control group	Hip prosthesis	Significance
Chromium	0.14 (0.07)	0.80 (0.36)	$P = 0.0089$
Cobalt	0.16 (0.05)	1.63 (0.91)	$P = 0.0073$

Biocompatibility of Cobalt–Chromium Alloys

Despite the *in vitro* evidence of good resistance to corrosion of cobalt–chromium alloys, there are concerns about these alloys used in orthopaedics. It is known, for instance, that cobalt and chromium ions can be released from implants. Corrosion has been associated with the formation of fibrous capsule around implants, a phenomenon that arises partly because of local inhibition of osteogenesis close to the implant surface. Osteolysis has also been found around cobalt–chromium hip implants, and this, too, has been shown to be associated with metallic debris and ions release as a result of corrosion.[28] the clinical consequence of these effects on bone is aseptic loosening of the implant.[29]

The association of aseptic loosening with abnormal bone resorption around the implant has led some workers to suggest that corrosion induces the release of cytokines that regulate bone metabolism.[30,31] Most of the cytokines involved in bone resorption are from the pro-infammatory group of cytokines. Two forms of interleukin 1, α and β, stimulate bone resorption and act synergistically with tumour necrosis factor-α (TNF-α) to induce osteoclasts to differentiate. TNF-α also causes collagen synthesis to be decreased and inhibits alkaline phosphatase activity in osteoblasts. Interleukin-6 stimulates bone resorption either by recruiting mature osteoblasts or by activating them.[32] All of these processes seem to be enhanced by the presence of chromium. For example, studies have demonstrated that peripheral blood mononuclear cells exposed to chromium extract show increased release of TNF-α and interleukin-6.[33] This suggests that they are likely to be induced to produce bone-resorbing cytokines, with consequent long-term and irreversible tissue damage.

A later study examined cobalt and chromium levels from blood samples of patients who had been diagnosed as suffering from aseptic loosening of the hip prosthesis.[34] When unstimulated peripheral blood mononuclear cells from patients with implant loosening were compared with those from a control group, it was found that the release of TNF-α was significantly higher, while that of interleukin-6 was significantly lower. As expected, these differences correlated with much higher levels of these metals in the serum levels of patients with implant loosening (see Table 4.2).

The mechanisms by which cobalt and chromium interfere with cytokine expression is not known, but there were differences found with the behaviour of peripheral blood mononuclear cells from patients suffering from implant loosening, and similar cells from the control group exposed to aqueous solutions of cobalt and chromium ions *in vitro*. This suggests that patients who suffer implant loosening do so because they display a hypersensitivity reaction to the metals. Such reactions have been reported clinically,[35,36] though not always associated with implant loosening. Other *in vitro* tests, notably on chemotactic activity of leukocytes, also suggest that a high proportion of patients with cobalt–chromium hip prostheses develop sensitivity to these metals, and that individuals who have a sensitivity to metals before receiving an implant run a high risk of having that sensitivity activated by the presence of a cobalt–chromium implant.[37]

Laboratory studies have been carried out to determine in detail how cobalt and chromium ions interact with cultured cells, in order to gain further understanding of their possible effects in the whole body. For example, in one study[28] solutions were prepared by electrochemical dissolution of cobalt–chromium–molybdenum orthopaedic alloy in 0.15 M NaCl, and then tested against cultures of human osteogenic bone marrow cells. Stock solutions of cobalt(VI) and chromium(III) chlorides were also tested in the same way. The solutions of corrosion products were found to have similar effects on cells to those of the stock solutions and generally there was a reduction in cell viability and of protein production, though this was less affected by the presence of metal ions than other physiological processes. ALP activity was also reduced, this being a significant finding because ALP is important in the initiation of mineralisation. Corrosion products from the alloy had a clear effect on human osteogenic cultures, with cells losing their smooth surface, and some dead cells being observed in peripheral areas.[38] Human cells have been found to be more sensitive to the presence of these ions than rat or rabbit cells, showing the importance of evaluating metal alloys against these cells, and not only in animal models, in order to determine their biocompatibility and the likely outcome of employing them under clinical conditions.

Problems have also been reported in connection with the toxic effect of cobalt and chromium against the immune system. One post-mortem study of patients who had been treated with cobalt–chromium replacement hip joints showed them to have experienced necrosis of the lymph nodes and changes in their T lymphocytes, both of which were demonstrated as being due to the effects of metal ions that arose from the prosthesis.[39] In another study, patients who had received cobalt–chromium hip implants showed a decrease in the prevalence of T lymphocytes;[40] results from other *in vitro* studies suggest that these effects are caused by the chromium rather than the cobalt, since this element is known to be the one which largely affects cell viability and diminishes DNA synthesis, as well as playing the primary role in inhibiting the release of cytokines.[41]

In some ways, more important than the local effects of cobalt and chromium ions are their systemic effects, and also the associated possibility that tumours may develop at sites remote from the implant. The ionic species that arise from corrosion of cobalt–chromium alloys give particular concern in this respect because hexavalent chromium is a known carcinogen in man[16] and cobalt is a possible carcinogen; this latter status reflects the fact that cobalt is a known mutagen and carcinogen in animals.[17] Clinical studies have shown that osteosarcomas can develop in patients who have cobalt–chromium hip implants, and that these are directly attributable to the prostheses,[42–44] Certainly, similar tumours have not been observed with either stainless steel or titanium alloy implants, and their occurrence in association with the use of cobalt–chromium alloys raises the question of just how suitable these alloys are for use as biomedical devices.

4 Stainless Steel

Stainless steel is the name given to a range of steels that include elements other than iron and carbon, which passivate the metal against corrosion. Typically, the passivating element is chromium which, under normal conditions, forms an adherent, tenacious oxide coating, through which water and oxygen cannot permeate, and hence which prevents corrosion. The main biomedical grade of stainless steel is called Type 316L and it is popular for use as osteosynthesis plates in orthopaedic applications.[45] This popularity arises from a satisfactory combination of good mechanical properties and acceptable cost.

Although type 316L stainless steel shows a relatively low corrosion rate under external conditions, within the body it will undergo corrosion at a more rapid rate.[46,47] This is because the natural surface oxide coating is not stable on exposure to physiological fluids, and both crevice and pitting corrosion are therefore able to take place within the body.[48] Even after times of implanation as short as two months there is clear evidence of crevice corrosion with this material.[49] This corrosion, or pitting corrosion, rarely occurs to such an extent that the implant fails mechanically. On the other hand, it necessarily occurs with release of ions into the surrounding tissues. For example, the *in vitro* corrosion of stainless steel in physiological solutions has been shown to release the toxic ions Cr^{6+}, Ni^{2+} and Mo^{6+}, as well as the more benign Fe^{3+}, and in a series of cytotoxicity tests on rat bone marrow stromal cells these ions were found to show toxic effects in the order $Cr > Mo = Fe > Co > Ni$. The presence of these ions in the body may cause local inflammation or other adverse reaction,[50] such as ion accumulation in the internal organs,[51] and can lead ultimately to implant loosening.[52]

Because of its biological importance, corrosion of 316L stainless steel has received considerable attention. It has been shown, for example, that both pitting and crevice corrosion are associated with the presence of sulfides in the metal,[53] a finding that has led to the specification of 0.01% as the maximum sulfur content in 316L stainless steel for biomedical applications.[54] The surface itself is also important in determining how well the implant will resist localised corrosion. For example, electropolishing improves corrosion resistance because it reduces the irregularity of the surface and may also remove embedded contaminants.[54]

The chemistry of the possible corrosion processes varies. Crevice corrosion is primarily influenced by the composition and structure of the adherent

Table 4.3 *Properties of annealed and cold-worked Type 316L stainless steel*

Property	Annealed	Cold-worked
0.2% Yield strength/MPa	503	945
Tensile strength/MPa	665	993
Elongation/%	41	13
ASTM grain size	7–8	9

surface coating, while pitting corrosion is influenced principally by the presence of non-metallic inclusions within the surface.[55] Electropolishing itself leaves behind a protective film, and this seems to be beneficial in improving the resistance of 316L stainless steel to crevice corrosion. However, subsequent treatments, such as bead blasting, can destroy this protective coating, and hence permit crevice corrosion to occur to the same extent as in untreated specimens. Consequently, great care is needed in the preparation of 316L stainless steel implants if their corrosion properties are to be satisfactory for their intended biomedical use.

Type 316L stainless steel bone fixture plates can be used in either the cold-worked or the annealed condition. The cold-worked material has higher strength, but is more difficult to form to bone contours under the conditions in an operating theatre. Some indications of how properties vary are shown in Table 4.3.

The behaviour of stainless steel within the body is complicated, and of varying degrees of acceptability. Within orthopaedics, it has frequently been found to exhibit interfacial corrosion defects.[56,57] It also develops defects during the handling and manufacturing stages, these being of various kinds, including scratches, drilling defects, and formation of metal tongues and splinters.[58] Stainless steel is also used in the construction of miniplates and screws designed for the stabilisation of bony fragments in the maxillofacial region. The use of such devices is now the method of choice in the fixation of mandibular fractures and other repairs to the bones of the face.[59]

The degradation of stainless steel within the body is due to a combination of electrochemical corrosion and wear. Corrosion results in loss of soluble ionic species, as already described, while wear gives rise to metal particles. These particles can accumulate in the tissues around the implant and cause discoloration.[60] They may also lead to a variety of other undesirable consequences, ranging in severity from mild fibrosis to infection and necrosis.

Studies of retrieved implants has given useful information on the behaviour of this metal under biological conditions. In one study of recovered plates and screws, 43 plates and 172 screws that had been implanted for times varying from six weeks to one year were examined by both light microscopy and SEM. Results are summarised in Table 4.4.

All of the recovered devices were found to have experienced mechanical damage, though the extent varied considerably in both distribution and severity. These devices had been handled with pliers and/or screwdrivers

Table 4.4 *Damage to recovered stainless steel plates and screws*

	Plates (43)	Screws (172)
Handling defects	100%	100%
Corrosion defects	19%	7%

during placement, and this had caused the observed damage. These defects were considerably worse than those observed on a similar set of recovered titanium plates and screws. The handling requirements of these devices are demanding, because of the need to fit them within the confined space and angled region of the mandible. Handling was not the only problem, since it was apparent that there had also been micro-motion and friction, so-called fretting, between the plates and the screws and this had led to mechanical degradation. Metal particles have been observed to accumulate in the region of such implants as a result of fretting conditions in service.[61]

Recovered miniplates seem to show no evidence of corrosion on their free surfaces,[62] whereas recovered orthopaedic devices have been found to exhibit pitting corrosion on their free surfaces.[63] This has been attributed to the repassivation mechanisms of this material, that are able to prevent gross breakdown of the metal. However, crevice corrosion has been found on screws and plates, and is generally localised around the natural crevices on the screws, or formed between the plates and the screws due to imperfect fitting of the screws into the holes.

Despite these defects, patient satisfaction with these devices appears good, and in the study of recovered devices[63] none had been removed because of adverse clinical response. Instead, all were recovered on a routine basis. In some cases, bone was found to have grown into small spaces between the plates and the screws, demonstrating that despite the potential problems of corrosion, stainless steel actually shows good biocompatibility in these bone-contacting applications. This aspect of its performance is now considered in detail.

Biocompatibility of Stainless Steel

There are two aspects to the biocompatibility of stainless steel, namely the effect of the metal surface and the effect of corrosion products that are released *in vivo*. In terms of the surface chemistry, the interaction of osteoblasts with 316L stainless steel has been found to be very good, indicating that the surface of this material has good biocompatibility towards these cells.[64] For example, osteoblasts attached strongly when exposed to polished discs of the metal in cell cultures for time periods of between 1 and 21 days. Adhesion to 316L stainless steel has been found to be stronger than to glass, and is comparable to the adhesion of osteoblasts to the titanium alloy Ti-6Al-4V,[65,66] with cell proliferation actually being greater on stainless steel than on the titanium alloy.

There have been several reports on the adverse effects of corrosion products of 316L stainless steel on both organs and tissues of animal models. For example, chromium has been found to accumulate in the livers of mice, and this leads to ultra-structural changes in these organs.[67] Semeniferous cells of mice are also reported as being affected by corrosion products from stainless steel,[68] as were their spleens.[69] Cell culture studies have been carried out using cultured bone marrow cells, which can be kept in conditions that allow them to proliferate and differentiate. Cells of this type have been obtained from rat and from rabbit[70] and in both cases exposure to stainless steel corrosion products inhibited proliferation and impaired the functioning of these cells. This suggests that corrosion products from stainless steel will damage the function of bone into which an implant is placed, and thus lead in time to implant loosening.

One method of attempting to improve the biocompatibility of 316L stainless steel has been ion implantation using N^+ ions.[71] This technique employs an ion beam and causes a thin layer of ion-implanted metal to be formed as the surface of the implant. This altered surface has modified properties compared with its untreated parent; for example, it may show improved wear characteristics and reduced corrosion, both of which have the potential to improve the biocompatibility. The mechanism of the changes is not fully understood, but it is known that iron nitride precipitates are formed in the ion-implantation process,[72] and these precipitates are assumed to be responsible for the enhanced wear resistance. Other studies have demonstrated that, under high N^+ fluxes, stainless steel develops well-defined nitride phases such as α''-$Fe_{16}N_2$ and ε-Fe_2N_{1-x} within it.[69] The actual nature of these precipitates varies with the carbon content of the steel, so that it is a combination of ion dose and initial composition that determines the total amount and type of iron nitrides that are formed. Overall, however, the ion-implanted material shows only modest improvement in corrosion behaviour compared with the original 316L stainless steel, and it is likely that any improvement in its performance *in vivo* would result from its altered surface, which in turn affects the tribology, and also the nature of any interactions with calcium and phosphate ions in the surrounding tissues.[73]

5 Titanium and Its Alloys

Titanium is a transition metal that is readily able to form solid solutions with elements whose atoms lie within about 20% of the size of the diameter of titanium atoms. Titanium itself has a high melting temperature, 1678 °C, and exists in hexagonal close packed geometry, the so-called α structure up to 882.5 °C, and above this temperature it adopts a body centred cubic structure, the so-called β structure.[74] In alloys, titanium can exist in a variety of structures, based on either pure or near α or β forms, or mixtures of α and β forms. Alloying elements tend to fall into one of three categories, namely α-stabilisers, such aluminium, oxygen, nitrogen and carbon, β-stabilisers, generally metals, such as vanadium, iron, nickel and cobalt, and neutral, such

Table 4.5 *Properties of selected experimental and commercial orthopaedic alloys of titanium*

Alloy	Micro-structure	Elastic modulus/ GPa	Yield strength/ MPa	Ultimate tensile strength/MPa
cpTi	α	105	692	785
Ti-6Al-4V	α/β	110	850–900	960–970
Ti-6Al-7Nb	α/β	105	921	1024
Ti-5Al-2.5Fe	α/β	110	914	1033
Ti-13Nb-13Zr	α'/β	79	900	1030

as zirconium. This latter group has no effect on the stability of the respective phases of titanium.

For biomedical applications, α and near-α alloys are preferred, because their corrosion resistance is superior to that of other types of titanium alloy. Variations in the processing conditions are used to control other aspects of titanium alloys, such as micro-structure, which in turn affects other properties, such as ductility, strength, fatigue resistance and fracture toughness. Table 4.5 gives examples of titanium alloys, their phase structures, and the resulting physical properties.

A point worth noting is that, although the elastic moduli are much lower than those of other acceptable metal alloys, the value is still much higher than that of bone (10–30 GPa). Hence there are still problems of stress shielding and bone resorption around titanium-based orthopaedic implants.[75] These problems have resulted in a number of proposed solutions for more flexible designs and low modulus materials. Proposals have included the use of carbon–carbon and carbon–polymer composites, whose modulus can be tailored to more closely approximate to that of bone than metals.[76] However, there are major problems with such materials, which typically undergo severe degradation within the body and have extremely poor tribological characteristics.

In fact, the introduction of titanium alloys was a consequence of the search for lower modulus materials. Though not ideal, α/β titanium alloys do have elastic moduli about half that of stainless steel or cobalt–chromium alloys, and certain experimental alloys have even lower values. As a result, they do go some way towards addressing the problem of stress shielding in orthopaedic repair.

When in the body, orthopaedic implants are subject to cyclic loading as a result of the motion of walking. This causes a cyclical pattern of plastic deformation in those microscopic zones of stress concentration caused by scratches or other inhomogeneities in the alloy. Fatigue itself actually depends on a number of factors, such as implant shape, material, processing and type of cyclic load. Standardised fatigue tests are available for the evaluation of orthopaedic components, which include tension/compression loading, bending, torsion and rotating bending fatigue testing. In addition, joint simulator trials can be used in the later stages of the development of implants.

The response of an alloy, even well characterised materials such as Ti-6Al-4V, is extremely sensitive to the history of the specimen prior to testing, for example, heating and cooling regimes, which in turn affect features such as grain size, and α/β ratio and morphologies. Fatigue lifetime of Ti-6Al-4V can be influenced by carefully controlled shot peening, for example increasing it by some 10%.[77] This is a cold-working process in which the surface is impacted by small spheres which plastically deform the surface. This results in a residual surface compressive stress, which reduces the effects of other surface-related mechanisms such as fatigue or corrosion. Though shot peening will improve fatigue limit, it does so at the cost of increased surface roughness. Consequently, for optimal all round performance in the body, a balance has to be struck between the beneficial and adverse effects of the shot peening process.

In general, the fatigue properties of titanium alloys are very sensitive to the condition of their surfaces. For example, Ti-6Al-4V can have its fatigue strength reduced by 40% by the introduction of notches,[77] and other surface preparation techniques can reduce the fatigue strength by even greater amounts.

Surface Chemistry of Titanium and Its Alloys

The surface chemistry of cpTi and Ti-6Al-4V are similar in that both have surfaces composed mainly of the oxide TiO_2 which has beneficial biochemical properties.[78] It is about 4–6 nm thick[79] and contains at least two types of hydroxyl group, acidic and basic. These are chemisorbed onto the surface[80] and can interact in different ways with the surrounding fluids of the body. In addition, TiO_2 as a surface layer can lead to both conformational changes and the denaturing of proteins.[78]

The precise composition and structure of the TiO_2 film vary with the details of composition of the titanium alloy, and also with the thermal, mechanical and chemical history of the specimen. Surface contamination is considered an important variable,[81] because a newly formed titanium dioxide surface is very reactive. Failure of dental implants to osseointegrate has been correlated with the presence in the surface of contaminants such as iron, zinc, tin and lead, which seem to have inhibited acceptance of the surface by the adjacent osteoblasts. To overcome such problems, modern implants are subjected to a variety of pretreatments, aimed at altering their surfaces and improving their biocompatibility. The object of all this pretreatment is to achieve a desired state of surface condition, such as cleanliness, passivation and topography. This latter is important, because it influences cell adhesion, morphology and proliferation.[82] Polishing is often used on titanium-based dental implants to enhance gingival attachment. Nitric acid treatment can be used to homogenise the passive film, anodising can be used to prepare thicker, denser films, and various roughening techniques can be employed to provide surfaces capable of enhanced osseointegration.[83–85]

Detailed studies of the surface of various titanium alloys have been carried out, and consideration given to the biological significance of the findings.[86]

Table 4.6 *Variation in surface composition of Ti-6Al-4V with surface treatment, as determined by XPS*

Initial treatment	Passivation	Autoclaving	Basic TiOH (atom%)	Acidic OH (atom%)	O 1s (atom%)
Polishing	None	No	14.21	32.01	53.78
Polishing	None	Yes	10.76	21.66	67.58
Brazing, 2 h	None	No	12.58	27.59	59.83
Brazing, 2 h	None	Yes	12.70	17.19	70.11
Brazing 8 h	None	No	16.52	24.79	58.68
Polishing	Nitric acid	No	11.34	26.05	62.61
Polishing	400 °C in air	No	9.57	26.09	64.34
Polishing	Boiling water	No	14.34	21.46	64.20

X-Ray photoelectron spectroscopy, XPS, was used to determine the nature of the surfaces of cpTi, Ti-6Al-4V and Ti-6Al-7Nb, and the alloys were found to have different surfaces from that of cpTi. In particular, their surface oxide films also contained the alloying elements Al, V and Nb, and feature which is likely to influence the adsorption of proteins and their conformation on the surface. This, in turn, is expected to modify the metal–cell interaction.

Because of its importance in biocompatibility, and also corrosion and wear, there have been a variety of approaches to altering the details of the surface of titanium and its alloys. In a typical study, Ti-6Al-4V was subjected to a number of different treatments aimed at altering the surface and increasing the passivation.[87] The surface finishes studied were (i) polished, (ii) brazed at 970 °C for 2 hours and (iii) brazed at 970 °C for 8 hours. Passivation techniques employed were (i) nitric acid passivation, (ii) heating in air at 400 °C and (iii) ageing in boiling water. Some specimens were also left untreated at this stage, but of these groups of specimens, some were subsequently treated further by autoclaving in steam at 121 °C for 30 minutes. This gave a total of 24 different treatment regimes. Having prepared these different specimens, they were examined with high-resolution XPS. Selected results are shown in Table 4.6.

The results show that passivation at 400 °C gave alloys with the lowest content of suboxides and metallic elements in the surface, and also the thickest oxide coatings, regardless of the initial condition of the surface. However, this treatment seemed to be accompanied by desorption of hydroxyl groups in the hydration layer on the surface of the alloy. There was some doubt as to the suitability of passivation by thermal treatment at 400 °C for brazed materials, because traces of copper and nickel were found in their surfaces, and these are likely to be lost on corrosion within the body, and hence go on to create adverse reactions in the tissues. Other passivation methods did not give rise to these elements in the surface, and are therefore preferred for brazed titanium alloys.

The oxide surface on these alloys was found to vary in thickness from about 1 nm to about 4.2 nm, depending on treatment. It was thickest for thermally treated specimens, and thinnest for the polished as-received specimens. Given its protective function, this suggests that the additional passivating treatment is desirable in generating surfaces that protect the underlying alloy from corrosion and which interact favourably with the body to promote bone ingrowth and osseointegration.

In a related study, atomic force microscopy was used to complement results obtained by XPS.[88] Specimens were prepared similarly in terms of initial treatment and passivation, though none were subsequently exposed to steam autoclaving.

XPS revealed that aged oxide surfaces were highly hydrated and contained amphoteric hydroxyl groups. These hydroxyl groups were in both bridged and terminal geometries. Atomic force microscopy confirmed that these sites were critical in determining the biocompatibility and osseointegration behaviour of these materials. Specifically, these seem to be the sites at which proteins adsorb from the body fluid surrounding the newly placed implant, a process which is important in the early stages of interaction with the body leading to successful osseointegration.

The influence of surface composition on corrosion behaviour has been considered experimentally. Corrosion is important, as we have seen, because of the influence of its reaction products on tissues, which in turn strongly influences how the tissues respond to the presence of the implants. The cpTi corrodes to give mainly neutral species at pH 7, mainly $Ti(OH)_4$, with only traces of ionic species, typically less than 10^{-12} mol dm^{-3}.

Wear Behaviour of Titanium and Its Alloys

Natural healthy joints show excellent tribological characteristics as a result of the properties of the components, namely articular cartilage and synovial fluid.[89] Sadly, artificial joints are less satisfactory in this regard, their lower lubrication performance being a consequence of the higher rigidity of their component materials compared with their natural counterparts. In general, friction between artificial materials is greater than in natural joints, with the result that non-recoverable wear takes place. The topic of wear of UHMWPE components was dealt with at length in Chapter 2, where it was shown that the accumulation of wear debris had undesirable biological consequences, typically inflammation, osteolysis, pain and eventual loosening of the implant.

Although the main particulate debris is polyethylene, it arises as a result of the tribological interaction of the UHMWPE component with the metal components of the joint.[90] In general, the alloy Ti-6Al-4V forms a satisfactory association with UHMWPE when used together in total joint replacement. However, scratches on the femoral head can greatly increase any wear associated with the frictional contact of these two materials, and such scratches are usually distributed over the femoral head in a non-uniform manner.

The alloy Ti-6Al-4V tends to be associated with greater rates of wear of the UHMWPE acetabular cap than, for example, cobalt–chromium alloys. This has been explained in terms of the mechanical instability of the metal oxide layer which covers the surface of titanium alloy components. It has been suggested that this layer is disrupted as a result of normal or shear stresses. Titanium is very reactive, and disruption and removal of the oxide layer results in rapid reforming of more oxide layer in its place. Overall, this sets up a continuous cycle in which metal oxide is removed, further metal is converted to oxide, and then removed, and so on, leading to a gradual loss of alloy from the femoral head. High surface roughness will follow such loss of metal, which in turn enhances the rate of wear of the UHMWPE component. The lost material may act as a third body abrasive, still further increasing rate and extent of wear.

To overcome the relatively poor tribological characteristics of Ti-6Al-4V for use in articulating surfaces, a number of approaches have been tried in attempts to improve the surface hardness, and thus to minimise the generation of UHMWPE wear debris.[91] This include PVD coating to generate surface layers of TiN or TiC, ion implantation (with N^+ ions), thermal treatments,[92] or laser alloying with TiC.[93] Ion implantation has been the most widely studied treatment, but results have been mixed. Very little improvement has been reported in the sliding wear resistance of the alloy, though there has been an improvement in the resistance to abrasion.[91] Although surface treatments are known to produce a harder layer composed of various metal oxides and possessing improved lubrication, to date there have been insufficient data collected to confirm that long-term wear characteristics of such treated surfaces are better than those of untreated alloys.

Oxygen diffusion hardening has been used to improve the wear resistance of the alloy Ti-6Al-7Nb.[94] This treatment results in a layer some 50 μm of transformed alloy, this material having a significantly lower coefficient of friction against UHMWPE than the parent alloy. Similar treatment of the alloy Ti-6Al-2.5Fe gave comparably good results, with the modified material showing much lower friction against UHMWPE than the unmodified alloy.[95]

Sliding wear tests have been used to assess the tribological properties of the newer titanium alloys. These include Ti-15Mo-5Zr-3Al, Ti-12Mo-6Zr-2Fe and Ti-13Nb-13Zr where, as before, the numbers refer to the percentage of the metal in the alloy. In general, all of these alloys show improved friction and wear behaviour compared with Ti-6Al-4V. For example, Ti-12Mo-6Zr-2Fe showed a friction coefficient less than half that of Ti-6Al-4V in a pin-on-disk test against poly(methyl methacrylate) in distilled water.[96] At low loads and after 10^5 cycles, this alloy showed no change in surface roughness and no surface scratching. The enhanced wear due to the release of third body metal oxide particles that is associated with the use of Ti-6Al-4V was not seen with this newer alloy, and it was also found to exhibit a much lower friction coefficient with UHMWPE than Ti-6Al-4V.

Abrasive wear behaviour of Ti-13Nb-13Zr was improved considerably by diffusion/oxidation hardening. Final properties were directly comparable with those of cobalt–chromium–molybdenum alloy, and significantly better than those of Ti-6Al-4V or even of TiN-coated Ti-6Al-4V.[97] This hardening process operates by the creation of a new substrate from oxygen diffusing into the metal, rather than depositing a new coating, as is the case for nitrogen

Table 4.7 *Properties of as-received and nitrogen-diffusion modified Ti-6Al-4V*

Metal (Ti-6Al-4V)	Corrosion potential (in Hank's neutral salt solution)/mV	Vickers microhardness (10 g, 25 s)
As received	−165	160 ± 15
Nitogen-modified	−90	380 ± 55

implantation. It results in the formation of a thin blue ceramic coating some 0.2 μm thick and consisting of a mixture of TiO_2, TiO and ZrO_2, the latter component being especially beneficial in conferring improved wear resistance.

Surface Modification of Titanium and Its Alloys

Surface modification by nitrogen diffusion has been considered as one way of improving both the corrosion resistance and surface hardness of the alloy Ti-6Al-4V.[98] In a typical study, specimens were prepared by heating them at 566 °C for 8 hours in a nitrogen atmosphere, after which they were cooled to room temperature in an atmosphere of nitrogen or of argon. Electron microscopy revealed that these specimens underwent a considerable change in surface morphology as a result of this treatment, caused by the surface layer of TiO_2 undergoing reaction to yield a variety of compounds of the general type TiN_xO_y and TiN_x in the upper layers of the surface. Corrosion and hardness properties were altered by this treatment, as shown in Table 4.7.

These values illustrate the improvements in properties of nitride-coated titanium alloys prepared by the diffusion process. These surfaces are likely to offer other advantages, namely improved cell attachment and improved wear. Cell attachment is likely to improve because cells are known to favour rougher surfaces.[99] They are able to span surface irregularities, such as grooves, and to extend processes into them, increasing overall area of contact and enhancing mechanical interlocking.[99] Wear is likely to be improved because TiN_x surfaces are known to be less susceptible to wear than TiO_2 surfaces,[100] and also to show reduced coefficient of friction, both of which suggest that these nitride-containing surfaces are likely to show improved properties when used as hip prostheses.

To date, there have been few accounts of the performance of this type of modified alloy under clinical conditions. In one study, however, some four retrieved femoral heads were studied in detail, using a variety of techniques, and some preliminary conclusions drawn.[101] The retrieved components had been in place for time periods ranging from 18 months to 8 years, and all had been articulating against UHMWPE counterfaces. Two of the four components showed macroscopically intact TiN coatings after 6–8 years of service. The other two showed fretting and coating breakthrough, which seemed to have arisen due to third body wear. The additional hard particles needed to promote this may have arisen from delamination of surface asperities and may have been augmented by particles of poly(methyl methacrylate) from the bone cement. These results suggest that TiN-coated titanium alloy may not be capable of resisting third-body wear *in vivo*. They also raise concerns about the future use of nitrogen-modified titanium alloys as femoral components for orthopaedic devices.

Biocompatibility of Titanium and Its Alloys

In general, the biological acceptability of metals is a function of their corrosion behaviour, and for optimum effects alloys need to be carefully tailored to minimise adverse reactions. Metal ions released by an alloy can lead to local adverse tissue reactions or to allergic reactions. The latter are not dependent on dose, unlike the former, and hence are not dependent on the rate of corrosion, merely on the fact that it takes place at all.

On the other hand, local tissue reactions are dependent on dose, and hence are affected by the rate of corrosion of the implant. Titanium and its alloys, notably Ti-6Al-4V, are widely recognised as having good corrosion resistance,[102] though there is evidence of release of titanium into the tissues. However, it has been difficult to assess the extent or significance of this phenomenon, because a variety of animal models have been used, and also a variety of implant procedures and retrieval techniques have been employed.[103] In baboons and in rabbits, no statistical differences were found between titanium levels in tissues between animals which had implants and those without.[104-106] On the other hand, in rats there was evidence of increased concentrations of titanium in the spleen of those animals exposed to implants. This was assumed to be because of the spleen's function as an organ for the accumulation of metals, though it may not be due to any toxic effect of the titanium so much as to an immune response of the spleen. Some degeneration of the liver was also observed, however, and this did seem to be due to some kind of toxic effect.[103]

There have been several studies published which have demonstrated that both cpTi and Ti-6Al-4V undergo ready osseointegration. There is direct contact between the living bone and the metal, with no fibrous capsule formation or other adverse development of the bone itself, at least as viewed with light microscopy.[107,108] Both cpTi and Ti-6Al-4V appear to have bioactive properties in the presence of tissue, allowing the direct contact of bone with the metal surface. This contrasts with the behaviour of 316L stainless steel and cobalt–chromium alloys, where there is little or no direct contact of fully differentiated bone with the implant surface.[109]

The nature of the surface oxide layer would seem to be likely to influence the quality of osseointegration of a particular implant, and this has been studied by Larsson *et al.*[110] The principal attributes they examined were oxide thickness and surface topography. In their study, Larsson *et al.* employed implants of a threaded design (3.75 mm diameter × 4.0 mm length) that were manufactured from cpTi rod. samples were ultrasonically cleaned for 10 minutes each in trichloroethylene, acetone and methanol, then either left unaltered (the control group), anodised, electropolished or electropolished

Table 4.8 *Surface variations in titanium implants (from Larsson et al.*[110])

Preparation	Oxide thickness/nm	Surface roughness/nm	Bone area (%)
Control (machined)	3–5	30.3	74.0
Anodised	180–200	40.8	76.3
Electropolished	2–3	2.9	73.6
Electropolished + anodized	180–200	32.3	67.7

then anodised. Resulting oxide thicknesses and surface roughness are shown in Table 4.8.

The anodised surfaces became a greyish purple colour due to the effects of light interference in the thick oxide coating that was formed. Following preparation, surfaces were further cleaned in the ultrasonic bath in 70% ethanol (3×10 minutes), then sterilised by autoclaving at 120 °C for 45 minutes. Implants were placed in adult New Zealand White female rabbits, each animal receiving one implant of each type, placed bilaterally in the tibial bones, using a careful surgical technique with generous irrigation with saline and low-speed drilling. After 1 year, the animals were sacrificed and the implants and surrounding bone tissue removed *en bloc* and examined microscopically.

In all cases, the implant site was mainly occupied by cortical bone, though some of the implant extended into the bone marrow cavity. No differences were found between the machined and electropolished groups, and though the group that had been electropolished and anodised showed reduced bone area, the effect was slight and insignificant. This was in contrast to results from short-term experiments, which have shown that thicker oxide coatings and greater roughness favour bone ingrowth and enhanced osseointegration.[111,112] Similar results have been obtained when the oxide layer has been enhanced by heat treatment in air, and the resulting implants placed into rats.[113] Implants of this type were found to require greater torques for removal.

The reason for the different findings in longer term experiments was not clear but two suggestions were made which might explain the results:

(1) That the electropolished, non-anodised surface acquires a thicker oxide coat *in vivo* as a result of exposure to the internal environment of the body. This is supported by results from a retrieval study, in which titanium-based implants collected from patients after varying times of exposure showed relatively thick surface oxide layers, and these layers were augmented with ions, mainly calcium and phosphorus, from the physiological environment.[114]

(2) That the rate of bone formation around these implants is influenced by ion release rates. Titanium ions have an inhibitory effect on calcification *in vivo*,[115] but release rate falls over time, due to self-passivation.[116,117] Absolute release rates are low, being of the order of ppm per cm^2 after several months in test solutions, but are assumed to depend on the properties of the oxide layer such as thickness, morphology, crystallinity and defect density. Although ion release has not been measured for the implants studied by Larrson *et al.*[108]

with the controlled surfaces, the possibility remains that ion release from the electropolished surfaces is higher than from the anodised surfaces, and that the slower rate of formation of bone around the electropolished surfaces is a result of this higher release rate and associated inhibitory effect.

Cell types vary in their responses to modified titanium surfaces. A number of studies have been carried out using cells such as macrophages,[118,119] fibroblasts,[120] periodontal cells,[121] epithelial cells,[122,123] osteoblasts,[124,125] and chondrocytes.[126] Osteoblasts have initially a greater attachment to rough, sandblasted surfaces with an irregular morphology, though it seems that surface roughness is not a good predictor of cell attachment and spreading *in vitro*. There is evidence that cell proliferation and differentiation, as well as matrix production, are altered by surface roughness, and that cells at different stages of differentiation respond differently to the same surface.[127] Hence it is probable that the outcome of the interaction between artificial materials and bone *in vivo* varies at early and late time periods, depending on the type and maturity of the cells involved. It is also likely that artificial materials are recognised in different ways by cells depending on the nature and conformation of specific proteins adsorbed on the surface. The amount and conformation would be expected to be influenced by the surface topography, which would thus indirectly influence the proliferation and differentiation of the cells.

Despite the fact that in general titanium alloys show good osseointegration, there are elements in the alloys Ti-6Al-4V and Ti-6Al-7Nb that are likely to give rise to more damaging ionic species. This is particularly true of vanadium, for which negatively charged corrosion products predominate. Vanadium itself is strongly cytotoxic,[128] it stimulates the production of mucus in the respiratory system,[129] and it is harmful to the production of blood.[130] On the other hand, the element vanadium is relatively less concentrated in the surface than in the bulk of the alloy Ti-6Al-4V, and hence there is less of it available for release into the surrounding tissues than might otherwise be the case.

Patients who have received Ti-6Al-4V implants have been found to excrete elevated levels of all three of these metals.[131] Using baboons as the animal model, details of the transport of the minor metals through the body have been established.[132] Vanadium, like titanium, has been shown to build up in the lung tissue, whereas aluminium became concentrated in the surrounding muscles, lungs and regional lymph nodes.

Among the elements used in titanium-based alloys, not only vanadium but also nickel and chromium are classified as toxic,[133] whereas titanium itself, and also zirconium and niobium, show little or no adverse effects when released under biological conditions. In fact, these elements are released in only low concentrations, if at all, due to their very stable corrosion products, which tend to be essentially insoluble oxides and which cannot be transported around the body with any degree of ease. Attempts to improve the resistance of Ti-6Al-4V to body fluids, and hence to limit the release of these elements, included the use of nitric acid passivation. However, this was not a success, since it led to an increase rather than a decrease in the levels of Al and V, as well as Ti, in serum-containing culture medium compared with untreated Ti-6Al-4V.[134,135]

It was in order to improve the biocompatibility of titanium-based alloys that Ti-6Al-7Nb and Ti-5Al-2.5Fe were first prepared and evaluated as biomaterials. The concept in both cases was to eliminate vanadium, because of its toxicity,[136] and to replace it with more benign elements. However, these alloys still contain aluminium, an element known to cause deficiencies in bone mineralisation[137] and neural disorders.[138,139] Further development work on titanium alloys has led to the elimination of this element, for example in Ti-12Mo-6Zr-2Fe. Conversely, there is evidence that large quantities of molybdenum are also detrimental to the biological acceptability of metals for implantation. This, in turn, led to the development of molybdenum-free alloys, such as Ti-15Zr-4Nb-2Ta-0.2Pd,[140] though the presence of palladium is also undesirable from the point of view of biocompatibility. The more recent development of Ti-13Nb-13Zr[141] may have solved the problem of biocompatibility of these alloys, since none of the elements present is associated with serious problems of adverse local reactions in tissues.

Vanadium in Ti-6Al-4V can be replaced with niobium because the elements have similar properties, at least in terms of their effect on stabilising the β phase of titanium. The final alloy in both cases contains both α and β titanium phases in similar ratios.[142] On the other hand, niobium is very much less cytotoxic than vanadium, so that any corrosion of the alloy Ti-6Al-7Nb will give rise to less toxicity than does corrosion of Ti-6Al-4V. In fact the niobium-based alloy actually undergoes corrosion much less readily than Ti-6Al-4V, from which we might anticipate that it will show generally improved biocompatibility. Certainly, this alloy is becoming more widely used in orthopaedics and it has also been considered for use as a casting alloy in dentistry. Its superior corrosion properties in models of saliva, notably dilute lactic acid, have been claimed to make it very promising for this application.[143]

The alloy Ti-6Al-2.5Fe has also received attention as a possible metal for the fabrication of biomedical devices.[144] It has been found to be as cytocompatible as Ti-6Al-4V and surface treatments, such as nitriding by plasma diffusion, have been shown to give identical outcomes in terms of improved biocompatibility for both alloys. The corrosion of both alloys was similar, but the alloy Ti-6Al-2.5Fe had the advantage that corrosion was not accompanied by release of vanadium.

One important way in which the environment inside the body differs from that which has usually been used in the *in vitro* study of corrosion and wear is that proteins are present. Their effect has received relatively scant attention, yet proteins can influence the corrosion of implant alloys by interacting with the electrical double layer that is established on the surface of the metal in an aqueous environment.[145,146] They can also influence wear behaviour, because they have a lubricating effect. The joint corrosion and wear behaviour of titanium alloys in phosphate buffered saline has been studied with and without the addition of two proteins, namely bovine albumin and foetal calf serum.[147] Three alloys were tested, *i.e.* Ti-6Al-4V, Ti-6Al-7Nb and Ti-13Nb-13Zr, with a variety of conditions involving both corrosion and wear. The study showed that Ti-13Nb-13Zr was more susceptible to wear than the other alloys, and that for all three alloys, the presence of protein reduced the extent of wear. Corrosion behaviour of the alloys varied considerably. For example, in the presence of protein, the corrosion resistance of Ti-13Nb-13Zr and Ti-6Al-7Nb was reduced, whereas the corrosion resistance of Ti-6Al-4V increased. Among other findings, this study showed that proteins have a profound effect in modifying the behaviour of titanium alloys in tests of those aspects that are important in determining *in vivo* durability.

Clinical Applications of Titanium and Its Alloys

Titanium and its alloys find widespread use in bone-contact applications, both in orthopaedics, as components of artificial hip and knee joints, and in dentistry, as implants for the support of either single crowns and a series of artificial teeth, so called bridges. The biology of the interaction of titanium-based alloys with bone has been studied extensively, and it has been established that bone formation around the implant derives from the adjacent bone bed, and involves growth towards the implant.[148] The surfaces of these titanium-based implants, though, are highly attractive substrates for bone cells, and differentiating osteoblasts have been shown to be capable of laying down a mineralised collagen-free matrix in direct contact with the surface oxide layer.[149,150] Other titanium surfaces, by contrast, influence the surrounding bone to lay down unmineralised tissue.[151] This is generally thin and somewhat variable, and often associated with zones of mineralisation.[152]

Effect of Titanium Release into the Body

The current clinical trends are towards longer implantation times and increased surface areas.[153] These, in turn, make titanium ion release more likely, and studies of the effect of such ion release more important. The earliest study of this topic was by Ferguson *et al.* in 1960,[154] who examined systemic titanium levels in rabbit following the placement of a titanium based implant. They measured the amount of titanium in samples of the liver, spleen, kidney, lung and muscle, and found that in the majority of animals there was no difference between titanium levels with or without implants. However, in a minority of cases, elevated levels of titanium were detected in the spleen and lung.

Much later, as we have seen, baboons were used as experimental animals, and ion release from Ti-6Al-4V was studied,[155] where this alloy had been used to fabricate prosthetic segmental devices for placement in the long bones. Titanium concentrations in the serum were found not to vary between experimental and control animals, but levels in the urine were six times greater in the experimental group compared to the controls, as were levels in the lung, spleen and regional lymph nodes.

A more recent study returned to the use of rabbits as the animal model.[151] This used a very careful experimental protocol, in which disc-shaped implants (96.35 mm diameter × 2.3 mm thickness) were implanted into adult male rabbits. Groups of seven rabbits received implants for 1 month, 4 months and

Table 4.9 *Tissue titanium levels*[151]

Group	Ti content in lung (ng g^{-1})		Ti content in spleen (ng g^{-1})	
Control	25.9	4.54	77.0	24.5
Sham	25.0	3.01	94.3	21.2
1 month implant	24.6	1.68	41.8	9.9
4 months implant	24.8	1.76	56.7	13.0
12 months implant	24.3	1.00	90.1	27.6

12 months. Another group of seven acted as control, and a further group were the 'sham' group, *i.e.* were operated on, but no implant was placed in them. Titanium levels were determined using electrothermal atomic absorption spectrophotometry, following microwave digestion of tissue samples. Results are shown in Table 4.9.

Differences in titanium content of lung tissue were not statistically significant, and each group contained some specimens with titanium levels below the limit of detection. For the spleen, certain differences were significant, but the differences between the control, sham and 12 months implant groups were not statistically significant. Overall, though there were significant differences for shorter implantation times, these were to values below those of the control, sham and 12 month groups, showing that titanium implants do not lead to elevations in metal content of the spleen in these animals. This study thus found no evidence for an increase in systemic titanium levels as a result of implantation.

Most of the studies of the interaction of titanium and its alloys report the effect on hard tissues. However, the adjacent soft tissues can also be affected. In one study of this topic, the soft tissue covering plates used for fracture fixation was analysed following routine removal of the plates 18 months after implantation.[156] Monoclonal antibodies were used to stain for specific cell types and automatic image analysis used to quantify cell numbers, and also cell distribution around the implant and the black titanium debris within the soft tissue.

In all sections studied, macrophages, fibroblasts, T lymphocytes and neutrophils were observed. Fibroblasts are known to be involved in normal wound healing processes and their presence around an implant was to be expected. However, the presence of macrophages was less expected, and was indicative of a chronic and sustained response to the implant. This may have arisen because of the steady release of debris from the implant resulting from frictional forces between the tissue and the implant.[157] The additional presence of the T lymphocytes suggests that the response is a chronic granulomous inflammation reaction, a process which appeared to be still continuing 18 months after implantation.

Not all the patients in the study were found to have visible clumps of debris within the soft tissue. However, the authors recognised that debris below 1 μm in diameter would not be resolvable with the light microscope, and hence the lack of *visible* debris was not taken to imply a complete lack of debris. Rather, all tissues were assumed to have some level of titanium within them. It seemed, too, that such debris was relatively passive once it had accumulated sufficiently to be visible as black particles. Smaller particle size debris seems to be more damaging, with greater likelihood of phagocytosis and serious tissue response.[158] There is also evidence that, at such small particle sizes, chemical composition is not important. The mere physical presence of particles, regardless of their composition, is sufficient to stimulate a response.

Use of Titanium and Its Alloys as Dental Prostheses

The main biomedical uses of titanium and its alloys are in bone-contact applications, *i.e.* in orthopaedics and in implant dentistry. These applications exploit the excellent biocompatibility of these alloys when placed in contact with bone. In the case of implants for dentistry, their usual design requires them to partly protrude from the bone and through the soft gingival tissue. Clinical requirements are demanding: there has to be adequate bone for anchorage, and patients must have good blood supply to the implant site, practise good oral hygiene and be non-smokers, the latter because of the effect smoking has on reducing blood flow adjacent to the implant.[159]

As well as wrought titanium alloys, titanium has also received some attention for use in casting alloys for the preparation of dental prostheses. Here, it is the low density and low modulus that are the attractive features being exploited. The main metals used for dental castings are gold alloys and cobalt–chromium. However, there are occasions when these are not acceptable. This can be for a variety of reasons, including the heavy weight of removable dentures and/or hypersensitivity and allergic reactions to the constituent elements that are experienced by certain patients.[160,161] Cast titanium and titanium alloys have therefore been used to make a variety of dental prostheses, including crowns,[162] cast plate dentures[163] and frameworks for removable partial dentures.[164]

In order for these cast devices to perform satisfactorily in service, it is essential that the casting process is carried out carefully, and that porosity is avoided.[165] A particular problem with these castings is that titanium is a reactive element and will react with certain elements in the investment materials, notably with silicate-based or phosphate-bonded investments.[166] This produces hard, brittle surface layers on the casting.[167] In order to overcome this, oxides that are more stable than the titanium oxides, such as CaO, MgO and ZrO, are included in the refractories.[168] These reduce the extent of any surface reaction of the titanium with the investment materials, though they do not eliminate it completely. They also reduce the surface roughening effect that conventional investment materials have on titanium alloy surfaces.

In clinical practice, the residual surface layers arising from reaction with the investment material need to be removed, because they not only reduce the ductility and fatigue resistance of partial denture frameworks and clasps, but

Table 4.10 *XPS data for titanium plates*

Element	Contaminated surface	Cleaned surface (with nylon brush and detergent)	Cleaned surface (as before + sterilised with Ar plasma)
Ti	0.00	20.01	17.59
O	17.67	45.13	43.22
C	68.41	31.39	37.86
N	13.90	3.46	1.34

also impair polishability and may affect accuracy of fit of crowns and fixed partial dentures.[169] It is therefore necessary to treat the surfaces of cast titanium alloys by grinding and polishing them. This reduces or eliminates the undesirable surface oxide layer, and studies have suggested that such post-fabrication treatments are easier for cpTi or Ti-6Al-4V than for cobalt–chromium castings.[168]

Handling of Titanium-based Implant Devices

In using titanium-based implants, emphasis is often placed on maintaining them free of contamination, which is achieved by avoiding contact with other materials. The surface cleanliness of the implant is considered paramount,[170,171] and under clinical conditions precautions such as handling the implant with titanium-tipped forceps only are recommended. These precautions may be important in ensuring that an implant is successful. However, the effect of surface contamination by substances such as hydrocarbons on the osseointegration process is not known.

The surface chemistry of titanium-based model implants has been studied using X-ray photoelectron spectroscopy, XPS, and time-of-flight secondary ion mass spectroscopy, ToF-SIMS.[172,173] The models were plates of surface area 3 cm × 1 cm, cut from machined titanium pieces that had been manufactured by the usual techniques for proper implants, and these were examined by XPS and ToF-SIMS using MgKα X-rays. These plates were found to have considerable amounts of surface contamination, with high levels of carbon and no detectable titanium. The nature of this carbon was determined by fitting the C 1s core line, and also from SIMS analysis, and was shown to be rich in C–O functional groups and to contain, *inter alia*, fatty acids and amides.

Various cleaning techniques were applied to these titanium plates, and their effectiveness was determined, again by XPS and ToF-SIMS. Glow discharge argon plasma sterilising treatment was found to be effective at removing the organic contamination layer, and other techniques, such as simple washing with detergent and scrubbing with a nylon brush, followed by argon plasma treatment, were shown to give an acceptable surface even following deliberate bacterial contamination. Results are shown in Table 4.10.

Although the authors demonstrated that such surfaces may be returned to their original state by simple procedures, they were unable to comment on whether or not such procedures had a positive influence on osseointegration and long-term integrity of implants. Indeed, the fact that 'as-received' plates had such high levels of contamination by organic species that no surface titanium could be detected raises important questions about the nature of the surface on real implants, and also about our understanding of surface reactions in the immediate aftermath of implantation.

6 Precious Metals and Their Alloys

Precious metals, of which gold is the most important, are rarely if ever used as biomedical materials, because of their relatively poor mechanical properties. On the other hand, the chemical inertness has proved desirable, notably in dentistry, and here precious metal alloys are widely used. Gold, in particular, been used in dentistry, its relatively poor mechanical properties being overcome by using it the form of an alloy, typically with copper.[174] Indeed, the gold–copper system has been widely studied, and is the basis of the majority of gold-based alloys used in clinical dentistry.[175] The major use of these alloys is in casting components such as dental crowns, which are retained on the prepared remnant of a tooth with the aid of so-called luting cements, and which function as replacement structures for lost or damaged teeth.[176] As such, they have to withstand the twin assaults of the moist, corrosive environment of the mouth, and the cyclical compressive forces of biting and chewing.

For many years, these alloys were classified on the basis of their progressively decreasing gold content as Type I (highest gold content, greatest ductility) through to Type IV (lowest gold content, least ductility). More recently, this classification has been superseded by one which is more inclusive, in that it does not rely for definitions on the gold content, and can therefore be readily applied to alloys containing other noble metals, such as platinum or palladium. Current definitions are:

(i) High-noble: Containing at least 60 wt% noble metal, or which at least 40 wt% is gold;
(ii) Noble: Containing at least 25 wt% noble metal, with no specification of the gold content;
(iii) Predominantly base metal, with a noble metal content less than 25 wt%.

Some typical compositions of noble metal casting alloys for dentistry are given in Table 4.11.

As mentioned earlier, the mouth is a difficult environment in which to employ metals, because of the presence of saliva. This naturally secreted fluid, though varying in composition between individuals, and within the same individual depending on *inter alia* state of health, is an electrolyte solution, and contains a number of ions, notably chloride, that are highly corrosive. The mouth can also be acidic, as when soft drinks containing citric or phosphoric acid are consumed, and this also contributes to the risk of corrosion of metals.[177]

Table 4.11 *Typical compositions (wt%) of noble metal casting alloys*

Type	Ag	Au	Cu	Pd	Pt	Zn	Other
High-noble	11.5	78.1	–	–	9.9	–	Ir
	10.0	76.0	10.5	2.4	0.1	1.0	Ru
	25.0	56.0	1.8	5.0	0.4	1.7	Ir
Noble	47.0	40.0	7.5	4.0	–	1.5	Ir
	38.7	20.0	–	21.0	–	3.8	In 16.5
	–	2.0	10.0	77.0	–	–	Ga 7
	70.0	–	–	25.0	–	2.0	In3

In general, it is well established that certain metals are relatively resistant to corrosion, and that this is a consequence of the formation of a thin, compact layer of oxides, of the order of 3 nm thick, on their surface.[178] Similar studies on dental alloys have shown that similar layers will form on them, but that they vary in composition and thickness, and hence in the degree of protection they afford to the underlying metal.[179] In particular, studies have confirmed that copper and silver reduce the corrosion resistance of gold-based alloys,[180,181] though whether to an extent that impairs their clinical performance is not clear. Certainly, alloys for the fabrication of crowns are generally of the highest noble metal content consistent with the need for good mechanical strength and resistance to distortion under the loading conditions within the mouth.

One issue related to corrosion in the mouth is the effect of the presence of implants. The proximity of two dissimilar metals, *e.g.* a titanium-based implant and a gold-based crown, together with the necessary electrolyte solution might be expected to lead to significant corrosion of the titanium structure. However, in an *in vitro* study, Reclaru and Meyer monitored the galvanic current for a variety of titanium/noble metal alloy combinations.[182] Though they were able to detect such currents with all combinations tested, the actual magnitudes were small, and they concluded that titanium/gold based alloys caused negligible corrosion, and also that they showed no risk of triggering crevice corrosion. They came to similar conclusions for the titanium/palladium system, which suggests that the current noble metal alloys used in dentistry pose little or no threat to any adjacent titanium implants, regardless of their precise composition, provided the overall noble metal content is high.

Gold is also used in dentistry for the fabrication of orthodontic wires. Though not strictly a biomaterial when used in this way, corrosion can lead to an increase in systemic levels of metal ions, with potentially adverse results. As with gold-based crowns, orthodontic wires contain substantial amounts of copper, typically in the range 10–20% The corrosion resistance is generally considered satisfactory, though clearly the presence of the copper reduces the nobility of the alloy and imparts some tendency to corrode.[183]

The resistance of gold to corrosion[184] suggests it may be suitable for for other biomedical applications, for example as implants. Gold is also known to

Table 4.12 *Total bone area (%) after 1 month and 6 months*

Metal surface	Bone area, 1 month	Bone area, 6 months
Gold	33	45
Zirconium	56	57
Titanium	51	59

differ from most metals, including those used to fabricate implants, in that it does not form surface oxide layers, and hence presents a purely metallic surface to the adjacent tissue. The interface between gold implants and cortical bone of rabbits has been studied,[185] and the consequences of this underlying chemistry on the interaction with bone determined.

In this work, the implant–tissue interface for a number of combinations of rabbit cortical bone and metal (gold, zirconium and titanium) was examined using light microscopic morphometry and ultrastructural electron microscopy, following placement as screw-shaped implants in adult New Zealand white rabbits. In addition, the surfaces of the retrieved metal specimens were examined by Auger electron spectroscopy, AES. Specimens were retrieved after 1 month and 6 months, and examined. Gold specimens seemed similar to those of zirconium and titanium, though there was some evidence that its surfaces were slightly less rough than those of the other two metals. AES revealed the gold to be coated with an adherent organic layer comprising major amounts of carbon, nitrogen and oxygen, together with trace amounts of sulfur, chlorine, sodium and tin. Despite the presence of these substances, gold itself was still detectable, showing that what was present was of molecular dimensions, rather than a thick, macroscopic organic film. Gold was actually found to have a higher affinity for binding these organic substances than the other two metals, a result that was attributed to the fact that its surface atoms are effectively unsaturated through not being able to form an oxide, unlike zirconium or titanium. Similar results have also been obtained for the adsorption of albumin and other proteins,[186] where gold again showed greater levels of biomolecule binding than oxide covered metals, including titanium.

Examination of the interface revealed that bone grew into direct contact, with the extent varying between the different metal surfaces, as shown in Table 4.12. There was not only a much lower bone area in contact with the gold surfaces; what bone there was proved to be relatively poorly mineralised compared with that adjacent to zirconium or titanium. Hence, despite the greater affinity of the gold for deposition of biomolecules, the overall findings were that this surface is of inferior bone biocompatibility compared with zirconium and titanium. These results suggest that either the ZrO_2 and TiO_2 surfaces have a positive effect on the development of adjacent bone, or that the pure gold surface has a negative effect. The relative chemical inertness of gold suggests the former explanation, but work is still underway to determine which of these possibilities is the underlying reason for these observations.

7 Dental Amalgam

Amalgam seems first to have been first used for the restoration of teeth in the early part of the 19th century in Europe. It was just one type of metallic restoration: others included hammered gold leaf or lead, the latter placed while molten. Right from the start, the use of amalgam was controversial. Mercury was known to be toxic and the technique of inserting amalgam seemed crude compared with the meticulous approach needed to place gold foil. Consequently the use of amalgam was considered unethical. In America, the dispute between those dentists who would use amalgam and those who would not became extremely polemical, leading to the so-called *amalgam wars*. In fact, there was an early professional body, the American Society of Dental Surgeons, whose express purpose was to unite ethical dentists (*i.e.* those refusing to use amalgam) against the 'unethical' ones. Later, many individuals were involved in helping to formulate safe and reliable amalgams for dental fillings.[187] One of the most notable was G.V. Black, whose *Manual of Operative Dentistry* published in 1896 established the mechanical principles for sound cavity design for use with these more satisfactory amalgams. Finally, in 1929, the American Dental Association adopted a specification for dental amalgam, which included the requirement that the material be tested under defined conditions. This was an important step in eliminating unsatisfactory products from the market.

To prepare dental amalgams, a powdered alloy consisting mainly of silver and tin is mixed with liquid mercury. The powder may be produced either by lathe cutting or by milling a cast ingot of the silver–tin alloy. The resulting particles are irregular in shape. Alternatively, the liquid alloy may be atomised and allowed to condense, a process which results in particles having an essentially spherical morphology. Alloys of both these types are used in clinical amalgams, as also are mixtures of lathe cut and spherical particles.[188]

In clinical use, amalgam alloy is mixed with mercury in a process known as *trituration*. Although formerly done by hand, possibly in the hands themselves with a rolling action, modern dental surgeries tend to be equipped with vibratory mixers, and the unmixed amalgam is prepared by the manufacturers in two chambers of a small capsule. Immediately prior to mixing, the thin membrane that separates the alloy powder from the liquid mercury is broken, and the capsule inserted into the arm of the mechanical mixer and vibrated for the required length of time, typically 30 seconds, to bring about thorough mixing of powder and liquid. The freshly mixed amalgam, which has a plastic consistency, is then extruded from the capsule and into the cavity.

During the process of trituration, the surface layer of the silver–tin alloy dissolves in the liquid mercury, and there is a reaction that leads to the formation of new phases. These new phases are solid, and their formation causes the plastic amalgam paste to solidify. A number of metallurgical phases are involved in this transformation, details of which are given in Table 4.13.

The detailed metallurgy of the phases involved is complex, and changes in the silver content of the initial silver–tin alloy can lead to the formation of

Table 4.13 *Phases involved in the setting of dental amalgam*

Phase	Chemical formula
γ	Ag_3Sn
γ_1	Ag_2Hg_3
γ_2	$Sn_{7-8}Hg$
ε	Cu_3Sn
η	Cu_6Sn_5
Silver–copper eutectic	Ag–Cu

different phases which have correspondingly different physical properties. Silver–tin alloys anyway are brittle and difficult to grind uniformly unless a small amount of copper is included. This is limited to 4–5 wt%, since above this level the discrete compound Cu_3Sn is formed. Below this level, the presence of copper hardens and strengthens the Ag–Sn alloy.

Zinc may also be included in the alloy, typically at levels of around 1 wt%. The presence of zinc leads to amalgams that are less plastic than zinc-free amalgams, an important feature during finishing processes for fillings. The main purpose of adding the zinc, though, is for it to act as a scavenger for oxygen,[189] thereby reducing corrosion through minimising the occurrence of other metal oxides in the finished amalgam.

The main setting reaction of dental amalgam is as follows:[190]

$$Ag_3Sn + Hg \rightarrow Ag_2Hg_3 + Sn_{7-8}Hg$$

The final alloy also contains significant amounts of the unreacted γ phase, Ag_3Sn. Modern amalgams are formulated to include up to 30 wt% copper,[191] and this leads to a subsequent reaction, as follows:

$$Sn_{7-8}Hg + Cu \rightarrow Cu_3Sn + Hg$$

The elemental mercury that is formed in this reaction is then free to react with further silver–tin alloy, and form the desirable γ_1 phase. There are several advantages to these reactions occurring in the setting process of amalgams: resulting materials are less susceptible to creep and corrosion[192] and they reach their final levels of strength quicker than so-called conventional amalgams.[193] The absence of corrosion is regarded as particularly advantageous, because it eliminates the main route by which mercury can be released from the filling and enter the patient *via* the gastro-intestinal tract. As a result, high-copper amalgams are now the material of choice for the clinical repair of cavities,[194] and in certain countries, *e.g.* Germany, high-copper is the only type of amalgam that is permitted for clinical use.

Dental amalgam is used within clinical dentistry for a variety of *permanent* restorations, *i.e.* those designed to last several years, rather than merely weeks or months.[195] The actual survival time varies considerably, depending on the brand of material, condition being repaired, and patient factors such as age and quality of oral hygiene.[196] Amalgams are recommended for a range of cavities, including substantial ones needed to repair the molar teeth.

Amalgam itself has no adhesion to either dentine or enamel. This means that there is the potential for marginal leakage, especially with high-copper amalgams, which give rise to less corrosion product that might otherwise fill any marginal gap. Traditionally, cavity walls are coated with a layer of copal-ether varnish.[197] This results in a very thin layer of organic film, about 2 μm deep, which provides some modest sealing of the gap by flowing to fill any surface irregularities on the prepared wall.

More recently, amalgam restorations have been bonded in place using a bonding agent especially designed for the purpose, Using such materials, quite high experimental bond strengths have been recorded between bovine dentine and amalgam; it is, however, too soon to have demonstrated whether such bonding agents improve the longevity and clinical performance of amalgam restorations when used in human teeth.

Briefly, amalgam has the following advantages:

(i)   It is inexpensive;
(ii) It is strong and durable in the oral environment, and shows excellent (*i.e.* minimal) wear;
(iii) Its use is relatively insensitive to clinical technique;
(iv) It has a proven track record of over 150 years of clinical service.

On the other hand, it has the following disadvantages:

(i)   Lack of adhesion, which may lead to marginal leakage (see above);
(ii) It has to be retained mechanically, which in turn means it is not conservative of tooth structure (healthy tooth tissue has to be removed to create the necessary undercut cavity in order to retain the hardened material);
(iii) Its aesthetics are poor;
(iv) There is patient concern over toxicity.

Dental amalgam is a space filler, and its placement in the tooth causes a weakening effect. Consequently, techniques that are highly conservative of tooth tissue are generally employed by dentists and cavities are cut so as to avoid sharp angles because they also weaken the tooth by causing stresses to be concentrated at the corners.

Amalgam is prepared and placed under the driest possible conditions because if an unset amalgam comes into contact with a wet liquid, *i.e.* saliva or blood, the cavity margins become difficult to finish properly, and the adaptation is poor. In amalgams containing zinc, there is also the possibility of reaction of the metallic zinc with water to yield zinc oxide and hydrogen:

$$Zn + H_2O \rightarrow ZnO + H_2 \text{ (gas)}$$

The latter causes bubbling and expansion of the filling which, in severe cases, will result in pulpal pain and cuspal fracture.

When used properly, amalgams have extremely good properties as restorative materials being durable and showing little in the way of recurrent caries. Modern high-copper zinc-containing amalgams have been shown to have extremely good survival rates, typically being shown in one study to have over 90% surviving for at least 12 years.[198] The overall conclusion from studies of durability is that the lifetime of an amalgam restoration depends on three sets of factors:

(i)   the material (brand, composition, quality of mixing *etc.*);
(ii) the dentist (cavity design, condensation, moisture control *etc.*); and
(iii) the patient (oral health and hygiene, diet, occlusal forces applied, including possible tooth grinding during sleep, behaviour known clinically as bruxism).

The first two influence performance during the early part of the restoration's lifetime, whereas the latter emerge as important as the restoration ages.

Dental Amalgams and Health

Mercury is a toxic element, both as the free metal and in chemical combination.[199] Elemental mercury is relatively soluble in lipids, and is readily absorbed at the lung surface, where it is oxidised to Hg^{2+}. It is transported from the lungs by the red blood cells to other tissues, including the central nervous system. Mercury is readily methylated in the environment and, as methylmercury, easily crosses the blood–brain barrier and also the placenta into the foetus. Consequently, it may accumulate in the brain, and may also affect the unborn child. Inorganic and metallic mercury, by contrast, does not cross the blood–brain barrier, and hence from amalgam fillings in these forms it does not pose a threat to the brain.

Corrosion of dental amalgam fillings may occur under the conditions found in the mouth.[200,201] There is some inhibition of corrosion by the strong passivating layer of SnO on the γ_1 phase which, though soluble in acid solutions, is not under the relatively mild acidic conditions of active caries, *i.e.* about 4.9.[202] Despite this inhibition, some corrosion may occur, and this causes mercury to be released as ions which pass into the gastro-intestinal tract. However, the amount is limited, and there is no evidence that it is sufficient to cause any adverse effects.

There are many studies which show mercury to have negligible effect on the health and well-being of patients. For example, in a study in Sweden,[203] the possible effects of amalgam fillings were examined by evaluating the health of patients drawn from the ongoing Swedish Adoption/Twin Study of Ageing. The mean age of the subjects was 66, and the authors concluded that no negative effects on physical or mental health could be found from dental amalgam, even after controlling for age, gender, education and number of remaining teeth.

There have been numerous attempts to link the presence of dental amalgam with the disease multiple sclerosis, MS. There are difficulties in that MS is characterised by bouts of spontaneous but temporary remission, so that anecdotal accounts of improvements in health following removal of amalgam fillings are of no value in determining whether there is any relationship. In fact, what evidence there is shows there to be no relationship. Clausen[204] analysed the mercury content of brains from those who had suffered from MS in their lifetimes and compared the results with those from the brains of deceased non-sufferers. The overall levels showed no significant differences, but the lipid-soluble mercury levels were significantly *lower* in the MS sufferers. This was explained in terms of changes in both the blood–brain barrier and in vitamin B_{12} metabolism in those affected by MS. Whatever the explanation, it is clear that MS is not connected with increased levels of mercury in the brain, and any suggestion of a connection between dental amalgams and the disease seems to have no basis in fact.

Dentists and their assistants have a much higher level of exposure to mercury in the form of vapour than do patients,[205] yet studies have shown that there are no significant differences in health, mortality and morbidity compared with the general population. In fact, the only known and scientifically confirmed problem with mercury is the very rare instances of mercury hypersensitivity.[206] Studies have also been conducted on the health of children born to dental personnel, and again there appear to be no risks of abnormalities in neonates in this group.[207] In conclusion, the scientific evidence suggests that the use of amalgam fillings poses no threat to the health of the dental personnel carrying out treatment and that, once placed and set, there is similarly no evidence that amalgam poses a threat to the health of patients.

8 References

1 K. Bordji, J. Y. Jouzeau, D. Mainard, E. Payan, P. Netter, K. T. Rie, T. Stucky and M. Hage-Ali, *Biomaterials*, 1996, **17**, 929.
2 L. Linder and J. Lundskog, *Injury*, 1975, **6**, 277.
3 H. Enneus and U. Stenram, *Acta Orthop. Scand.*, 1965, **36**, 115.
4 M. Long and H. J. Rack, *Biomaterials*, 1998, **19**, 1621.
5 W. Bonfield, in *Biomaterials and Clinical Applications*, ed. A. Pizzoferrato, A. Ravaglioli and A. J. C. Lee, Elsevier, Amsterdam, 1987, pp. 13–19.
6 R. M. Pilliar and G. C. Weatherly, in *Critical Reviews in Biocompatibility*, ed. D. F. Williams, CRC Press, Boca Raton, FL, 1985, pp. 371–403.
7 A. Bartolozzi and J. Black, *Biomaterials*, 1985, **6**, 2.
8 L. D. Dorr, R. Blocbaum, J. Emmanuel and R. Meldrum, *Clin. Orthop. Rel. Res.*, 1990, **261**, 82.
9 R. Deutman, T. J. Rulder, R. Brian and J. P. Nater, *J. Bone Jt. Surg.*, 1977, **59A**, 862.
10 E. M. Evans, M. A. R. Freeman, A. J. Miller and B. Vernon-Roberts, *J. Bone Jt. Surg.*, 1974, **56B**, 626.
11 M. H. Huo, E. A. Salvati, J. Lieberman, F. Betts and M. Bansal, *Clin. Orthop. Rel. Res.*, 1992, **276**, 157.

12 D. F. Williams, *Fundamental Aspects of Biocompatibility*, Vol. 1, ed. D. F. Williams, CRC Press, Boca Raton, FL, 1981, Chapter 2.

13 R. van Noort, *J. Mater. Sci.*, 1987, **22**, 3801.

14 G. C. Clark and D. F. Williams, *J. Biomed. Mater. Res.*, 1982, **16**, 125.

15 J. McCann and B. N. Ames, *Proc. Nat. Acad. Sci.*, 1976, **73**, 950.

16 IARC Monographs on the evaluation of carcinogenic risks to humans: List of IARC evaluation, IARC, Lyon, 1996.

17 IARC Monographs on the evaluation of carcinogenic risks to humans: Vol. 52, Cobalt and cobalt compounds, IARC, Lyon, 1991.

18 E. Gocke, M. T. King, K. Ekhardt and D. Wild, *Mutation Res.*, 1981, **90**, 91.

19 T. G. Rossman, M. Molina, L. Meyer, P. Boone, C. B. Klein, Z. Wang, F. Li, W. C. Lin and P. L. Kinney, *Mutation Res.*, 1991, **260**, 349.

20 A. C. Mello-Filho and R. Meneghini, *Mutation Res.*, 1991, **251**, 109.

21 K. Takahashi, T. Imaeda and Y. Kawazoe, *Biochem. Biophys. Res. Comm.*, 1988, **157**, 1124.

22 S. Y. Liau, D. C. Read, W. J. Pugh, J. L. Furr and A. D. Russell, *Lett. Appl. Microbiol.*, 1997, **25**, 279.

23 Q. L. Feng, J. Wu, C. Q. Chen, F. Z. Cui, T. N. Kim and J. O. Kim, *J. Biomed. Mater. Res.*, 2000, **52**, 662.

24 M. Dusinska and D. Slamenova, *Biologia*, 1990, **45**, 211.

25 F. Denizeau and M. Marion, *Cell Biol. Toxicol.*, 1989, **5**, 15.

26 D. F. Williams, in *Biocompatibility of Clinical Implant Materials*, ed. D. F. Williams, CRC Press, Boca Raton, FL, 1981, Chapter 4.

27 D. C. Fricker and R. Shivanath, *Biomaterials*, 1990, **11**, 495.

28 H. Tomas, G. C. Carvalho, M. H. Fernandes, A. P. Freire and L. M. Abrantes, *J. Mater. Sci.; Mater. Med.*, 1996, **7**, 291.

29 J. Black and E. C. Maitin, *Biomaterials*, 1983, **4**, 160.

30 S. Wallach, L. V. Avioli, J. D. Feinblatt and J. H. Carstens Jr., *Calcif. Tissue Int.*, 1993, **53**, 293.

31 A. Pizzoferrato, S. Stea, L. Savarino, D. Grachi and G. Ciapetti, *Chir. Org. Mov.*, 1994, **79**, 245.

32 T. Kuroki, M. Shingu, Y. Koshihara and M. Nobunaga, *Br. J. Rheumatol.*, 1994, **33**, 224.

33 D. Granchi, E. Verri, G. Ciapetti, L. Savarno, A. Sudanese, M. Mieti, R. Rotini, D. Dallari, G. Zinghi and L. Montanaro, *Biomaterials*, 1998, **19**, 283.

34 D. Granchi, G. Ciapetti, S. Stea, L. Savarino, F. Filippini, A. Sudanese, G. Zinghi and L. Montanaro, *Biomaterials*, 1999, **20**, 1079.

35 M. D. Brown, E. A. Lockshin, E. A. Salvati and P. G. Bullough, *J. Bone Jt. Surg.*, 1977, **59–A**, 164.

36 A. Remes and D. F. Williams, *Biomaterials*, 1992, **13**, 731.

37 J. Black, *Host Response*, in *Orthopaedic Biomaterials in Research and Practice*, Churchill-Livingstone, New York, 1988, pp. 285–302.

38 H. Tomas, G. S. Carvalho, M. H. Fernandes, A. P. Freire and L. M. Abrantes, *J. Mater. Sci.; Mater. Med.*, 1997, **8**, 233.

39 C. P. Case, V. G. Langlamer, C. James, M. R. Palmer, A. J. Kemp, P. F. Heap and L. Solomon, *J. Bone. Jt. Surg.*, 1994, **76**, 701.

40 D. Granchi, G. Ciapetti, S. Stea, D. Cavedagna, N. Bettini, T. Bianco, G. Fontanesi and A. Pizzoferrato, *Chir. Org. Mov.*, 1995, **25**, 399.

41 J. Y. Wang, D. T. Tsukayama, B. H. Wicklund and R. B. Gustilo, *J. Biomed. Mater. Res.*, 1996, **32**, 655.

42   F. W. Sunderman, *Fund. Appl. Toxicol.*, 1989, **13**, 205.

43   J. J. Jacobs, D. H. Rosenbaum, R. M. Hay, S. Gitelis and J. Black, *J. Bone Jt. Surg.*, 1992, **74**, 740.

44   A. J. Aboulafia, K. Littleton, B. Smookler and M. M. Malawer, *Orthop. Rev.*, 1994, **23**, 427.

45   J. Beddoes and K. Bucci, *J. Mater. Sci.; Mater. Med.*, 1999, **10**, 389.

46   K. W. H. Seah and X. Chen, *Corros. Sci.*, 1993, **34**, 1841.

47   Y. Nakayama, T. Yamamuro, P. Kumar, K. Shimisu, Y. Kotoura, M. Oka, J. Kwafuku and T. Takashima, *J. Appl. Biomater.*, 1990, **1**, 307.

48   K. J. Bundy, *Crit. Rev. Biomed. Eng.*, 1994, **23**, 130.

49   M. Sivakumar, K. S. K. Dhanadurai, R. Rajeswari and R. Rhulasirman, *J. Mater. Sci. Lett.*, 1995, **14**, 351.

50   K. Merrit and S. A. Brown, *J. Biomed. Mater. Res.*, 1995, **29**, 627.

51   M. L. Pereira, A. M. Abrue, J. P. Sousa and G. S. Carvalho, *J. Mater. Sci.; Mater. Med.*, 1995, **6**, 523.

52   E. Leitao, R. A. Silva and M. A. Barbosa, *J. Mater. Sci.; Mater. Med.*, 1997, **8**, 365.

53   G. E. Eklund, *J. Electrochem. Soc.*, 1976, **123**, 170.

54   ASTM Standard F138–86, Standard specification for stainless steel bar and wire for surgical implants (special quality), 1986.

55   G. Hultquist and G. Leygraf, *Corrosion*, 1980, **36**, 126.

56   A. F. Harding, S. D. Cook, K. A. Thomas, C. L. Collins, R. J. Haddad and M. Milicic, *Clin. Orthop. Rel. Res.*, 1985, **195**, 261.

57   D. F. Williams and G. Meachim, *J. Biomed. Mater. Res., Symp.*, 1974, **5**, 1.

58   O. E. M. Pohler, in *Biomaterials in Reconstructive Surgery*, ed. L. R. Rubin, Mosby, London, 1983, pp. 158–228.

59   D. E. Altobelli in *Rigid Fixation of the Craniomaxillofacial Skeleton*, ed. M. J. Yaremchuk, J. S. Gruss and P. N. Manson, Butterworth-Heinemann, Boston, MA, 1992.

60   G. Meachim and R. B. Pedley, in *Fundamental Aspects of Biocompatibility*, Vol. 1, ed. D. F. Williams, CRC Press, Boca Raton, FL, 1981, pp. 107–144.

61   L. E. Moberg, A. Nordenram and O. Kjellman, *Int. J. Oral Maxillofac. Surg.*, 1989, **18**, 311.

62   S. D. Cook, K. A. Thomas, A. F. Harding, C. L. Collins, R. J. Haddad Jr, M. Milicic and W. L. Fischer, *Biomaterials*, 1987, **8**, 177.

63   S. D. Cook, E. A. Renz, R. L. Barrack, K. A. Thomas, A. F. Harding, R. J. Haddad and M. Milicic, *Clin. Orthop. Rel. Res.*, 1985, **194**, 236.

64   K. Anselme, B. Noel and P. Hardouin, *J. Mater. Sci.; Mater. Med.*, 1999, **10**, 815.

65   R. K. Sinha, F. Morris, S. A. Shah and R. S. Tuan, *Clin. Orthop. Rel. Res.*, 1994, **305**, 258.

66   C. R. Howlett, M. D. M. Evans, W. R. Walsh, G. Johnson and J. G. Steele, *Biomaterials*, 1994, **15**, 213.

67   M. L. Pereira, A. M. Abreu, J. P. Sousa and G. S. Carvalho, *J. Mater. Sci.; Mater. Med.*, 1995, **6**, 523.

68   M. L. Pereira, A. Silva, R. Tracana and G. S. Carvalho, *Cytobios.*, 1994, **77**, 73.

69   R. B. Tracana, M. L. Pereira, A. M. Abreu, J. P. Sousa and G. S. Carvalho, *J. Mater. Sci.; Mater. Med.*, 1995, **6**, 56.

70   S. Morais, J. P. Sousa, M. H. Fernandes, G. S. Carvalho, J. D. de Bruijn and C. A. van Blitterswijk, *Biomaterials*, 1998, **19**, 999.

71 E. Leitao, R. A. Silva and M. A. Barbosa, *J. Mater. Sci.; Mater. Med.*, 1997, **8**, 365.

72 B. Rauschenbach, *Nucl. Instrum. Methods Phys. Res.*, 1991, **B53**, 35.

73 C. A. P. T. Klein, P. Patka, J. G. C. Wolke, J. M. A. de Blieck-Hogervorst and K. de Groot, *Biomaterials*, 1994, **15**, 146.

74 E. W. Collings, *The Physical Metallurgy of Titanium Alloys*, ASM Series in Metal Processing, American Society for Metals, Cleveland, Metals Park, OH, 1984.

75 D. R. Sumner and J. O. Galante, *Clin. Orthop. Rel. Res.*, 1992, **274**, 124.

76 M. F. Semlitsch, H. Weber, R. M. Streicher and R. Schön, *Biomaterials*, 1992, **13**, 781.

77 A. W. Eberhardt, B. S. Kim, E. D. Rigney, G. L. Kutner and C. R. Harte, *J. Appl. Biomater.*, 1995, **6**, 171.

78 D. S. Sutherland, P. D. Forshaw, G. C. Allen, I. T. Brown and K. R. Williams, *Biomaterials*, 1993, **14**, 893.

79 J. Lausmaa, M. Ask, U. Rolander and B. Kasemo, *Mater. Res. Soc. Symp. Proc.*, 1989, **110**, 647.

80 K. E. Healy and P. Ducheyne, *Biomaterials*, 1992, **13**, 553.

81 A. P. Ameen, R. D. Short, R. Johns and G. Schwach, *Clin. Oral Implant Res.*, 1993, **4**, 144.

82 D. Buser, R. K. Schenk, S. Steinemann, J. P. Fiorellini, C. H. Fox and H. Stich, *J. Biomed. Mater. Res.*, 1991, **25**, 889.

83 B. D. Boyan, T. W. Hummert, D. D. Dean and Z. Schwartz, *Biomaterials*, 1996, **17**, 137.

84 Z. Schwartz, J. Y. Martin, D. D. Dean, J. Simpson, D. L. Cochran and B. D. Boyan, *J. Biomed. Mater. Res.*, 1996, **30**, 145.

85 M. Wong, J. Eulenberger, R. Schenk and E. Hunziker, *J. Biomed. Mater. Res.*, 1995, **29**, 1567.

86 C. Sittig, M. Textor, N. D. Spencer, M. Weiland and P.-H. Vallotton, *J. Mater. Sci.; Mater. Med.*, 1999, **10**, 35.

87 T. M. Lee, E. Chang and C. Y. Yang, *J. Mater. Sci.; Mater. Med.*, 1998, **9**, 439.

88 M. Browne, P. J. Gregson and R. H. West, *J. Mater. Sci.; Mater. Med.*, 1996, **7**, 323.

89 J. Fisher and D. Dowson, *Proc. Inst. Mech. Eng., Part H*, 1991, **205**, 73.

90 M. Jasty, *J. Appl. Biomater.*, 1993, **4**, 273.

91 H. A. McKellop and T. V. Rostlund, *J. Biomed. Mater. Res.*, 1990, **24**, 1413.

92 P. H. Morton and T. Bell, *Mém. Études Sci. Métallurgie*, 1989, **86**, 639.

93 C. Chengwei, Z. Zhiming, T. Xitang, Y. Wang and X. T. Sun, *Tribol. Trans.*, 1995, **38**, 875.

94 R. M. Streicher, H. Weber, R. Schön and M. Semlitsch, *Biomaterials*, 1991, **12**, 125.

95 J. Zwicker, U. Etzold and Th. Moser, in *Titanium '84 Science and Technology*, Vol. 2, Deutsche Gesellschaft für Metallkunde EV, Munich, 1985, p. 1343.

96 K. Wang, L. Gustavson and J. Dumbleton, in *Beta-titanium in the 1990s*, The Minerals, Metals and Materials Society, Warrendale, 1993, pp. 2697–2704.

97 A. K. Mishra, J. A. Davidson, P. Kovacs and R. A. Poggle, in *Beta-titanium in the 1990s*, The Minerals, Metals and Materials Society, Warrendale, 1993, pp. 61–72.

98 R. Venugopalan, M.A, George, J. J. Weiner and L. C. Lucas, *Biomaterials*, 1996, **20**, 1709.

99 J. Meyle, H. Wilburg and A. F. von Recum, *J. Biomat. Appl.*, 1993, **7**, 362.

100 A. Wisbey, P. J. Gregson and M. Tuke, *Biomaterials*, 1987, **8**, 477.

101 M. T. Raimondi and R. Pretrabissa, *Biomaterials*, 2000, **21**, 907.

102 T. M. Lee, E. Chang and C. Y. Yang, *J. Mater. Sci.; Mater. Med.*, 1999, **10**, 541.

103 D. Rodriguez, F. J. Gil, J. A. Planell, E. Jorge, L. Alvarez, R. Garcia, M. Larrea and A. Zapata, *J. Mater. Sci.; Mater. Med.*, 1999, **10**, 847.

104 J. L. Woodman, J. J. Jacobs, J. O. Galante and R. M. Urban, *J. Orthop. Res.*, 1984, **1**, 421.

105 S. J. Lugowski, D. C. Smith, A. D. McHugh and J. C. van Loon, *J. Biomed. Mater. Res.*, 1991, **25**, 1443.

106 P. D. Bianco, P. Ducheyne and J. M. Cuckler, *J. Mater.Sci.; Mater. Med.*, 1997, **8**, 525.

107 C. Johansson, J. Lausmaa, M. Ask, H.-A. Hansson and T. Albrektsson, *J. Biomed. Eng.*, 1989, **11**, 3.

108 J. W. McCutchen, J. P. Collier and M. B. Mayer, *Clin. Orthop. Rel. Res.*, 1990, **261**, 114.

109 T. Albrektsson, P.-I. Branemark, H. A. Hansson, B. Kasemo, K. Larsson, I. Lundstrom, D. H. McQueen and R. Skalak, *Ann. Biomed. Eng.*, 1983, **11**, 1.

110 C. Larsson, L. Emamuelsson, P. Thomsen, L. E. Ericson, B. O. Aronsson, B. Kasemo and J. Lasmaa, *J. Mater. Sci.; Mater. Med.*, 1997, **8**, 721.

111 C. Larsson, P. Thomsen, B. O. Aronsson, M. Rodahl, J. Lausmaa, B. Kasemo and L. E. Ericson, *Biomaterials*, 1994, **15**, 1062.

112 A. Wennerberg, T. Albrektsson and B. Andersson, *Int. J. Oral Maxillofac Impl.*, 1993, **8**, 622.

113 R. Hazan, R. Brener and U. Oron, *Biomaterials*, 1993, **14**, 570.

114 J. E. Sundgren, P. Bodo and I. Lundstrom, *J. Colloid Interface Sci.*, 1986, **110**, 9.

115 N. C. Blumenthal and V. Cosma, *J. Biomed. Mater. Res.*, 1989, **23**, 13.

116 K. E. Healey and P. Ducheyne, *J. Biomed. Mater. Res.*, 1992, **26**, 319.

117 K. E. Healey and P. Ducheyne, *J. Coloid Interface Sci.*, 1992, **150**, 404.

118 A. Rich and K. A. Harris, *J. Cell Sci.*, 1981, **50**, 1.

119 T. N. Salthouse, *J. Biomed. Mater. Res.*, 1984, **18**, 395.

120 P. Van Der Valk, A. W. J. Van Pelt, H. J. Busscher, H. P. D. Jong, C. R. H. Wildevar and J. Arenos, *J. Biomed. Mater. Res.*, 1983, **17**, 807.

121 D. Cochran, J. Simpson, H. Weber and D. Buser, *Int. J. Oral Maxillofac. Impl.*, 1994, **9**, 289.

122 B. Chehnoudi, T. R. L. Gould and D. M. Brunette, *J. Biomed. Mater. Res.*, 1989, **23**, 1067.

123 B. Chenoudi, T. R. L. Gould and D. M. Brunette, *J. Biomed. Mater. Res.*, 1990, **24**, 1202.

124 K. T. Bowers, J. C. Keller, B. A. Randolph, D. G. Wick and C. M. Michaels, *Int. J. Oral Maxillofac. Impl.*, 1992, **7**, 302.

125 J. Y. Martin, Z. Schwartz, T. W. Hummert, D. M. Schraub, J. Simpson, J. J. Lankford, D. D. Dean, D. L. Cochran and B. D. Boyan, *J. Biomed. Mater. Res.*, 1995, **29**, 389.

126 Z. Schwartz, J. Y. Martin, D. D. Dean, J. Simpson, D. L. Cochran and B. D. Boyan, *J. Biomed. Mater. Res.*, 1996, **30**, 145.

127 B. D. Boyan, T. W. Hummert, K. Kieswetter, D. Schraub, D. D. Dean and Z. Schwartz, *Cells Mater.*, 1995, **5**, 323.

128 R. C. Browne, *Br. J. Ind. Med.*, 1955, **12**, 57.

129 S. G. Sjoberg, *Acta Med. Scand.*, 1956, **154**, 381.

130 S. G. Sjoberg and K.-G. Rigner, *Nord. Hyg. Tidskr.*, 1956, **37**, 217.

131 J. J. Jacobs, A. K. Skipor, J. Black, R. M. Urban and J. O. Galante, *J. Bone Jt. Surg.*, 1991, **73A**, 1475.

132 J. L. Woodman, J. J. Jacobs, J. O. Galante and R. M. Urban, *J. Orthop. Res.*, 1984, **1**, 421.

133 S. G. Steinemann, *Titanium '84 Science and Technology*, Vol. 2, Deutsche Gesellschaft für Metallkunde EV, Munich, 1985, pp. 1373–1379.

134 B. F. Lowenberg, S. Lugowski, M. Chipman and J. E. Davies, *J. Mater. Sci.; Mater. Med.*, 1994, **5**, 467.

135 B. W. Callen, B. F. Lowenberg, S. Lugowski, R. N. S. Sodhi and J. E. Davies, *J. Biomed. Mater. Res.*, 1995, **29**, 279.

136 P. G. Laing, A. B. Ferguson Jr. and E. S. Hodge, *J. Biomed. Mater. Res.*, 1967, **1**, 135.

137 U. Meyer, D. H. Szulczewski, R. H. Barckhaus, M. Atkinson and D. B. Jones, *Biomaterials*, 1993, **14**, 917.

138 D. P. Perl and A. R. Brody, *Science*, 1980, **208**, 297.

139 P. R. Walker, J. LeBlanc and M. Sikorska, *Biochemistry*, 1989, **28**, 3911.

140 Y. Okazaki, Y. Ito, A. Ito and T. Tateishi, *J. Jpn. Inst. Metals*, 1993, **57**, 332.

141 P. Kovacs and J. A. Davidson, in *Titanium '92, Science and Technology*, The Minerals, Metals and Materials Society, Warrendale, 1993, pp. 2705–2712.

142 M. F. Semlitsch, H. Weber, R. M. Steicher and R. Schon, *Biomaterials*, 1992, **13**, 781.

143 E. Kobayashi, T. J. Wang, H. Doi, T. Yoneyama and H. Hamanaka, *J. Mater. Sci.; Mater. Med.*, 1998, **9**, 567.

144 K. Bordi, J. Y. Jouzeau, D. Mainard, E. Payan, P. Netter, K. T. Rie, T. Stucky and M. Hage-Ali, *Biomaterials*, 1996, **17**, 929.

145 S. R. Sousa and M. A. Barbosa, *Clin. Mater.*, 1993, **12**, 1.

146 D. F. Williams, in *Critical Reviews in Biocompatibility*, Vol. 1, ed. D. F. Williams, CRC Press, Boca Raton, FL, USA, 1985, Chapter 1.

147 M. A. Khan, R. L. Williams and D. F. Williams, *Biomaterials*, 1999, **20**, 631.

148 M. Weilander, *Dent. Clin. N. Am.*, 1991, **35**, 585.

149 J. E. Davies, R. Chernecky, B. Lowenberg and A. Shiga, *Cell Mater.*, 1991, **1**, 3.

150 B. Lowenberg, R. Chernecky, A. Shiga and J. E. Davies, *Cell Mater.*, 1991, **1**, 177.

151 L. Sennerby, L. E. Ericson, P. Thomsen, U. Lekholm and P. Astrand, *J. Dent. Res.*, 1992, **71**, 364.

152 A. Piattelli, A. Scarano, M. Piattelli and L. Calabrese, *Biomaterials*, 1996, **17**, 1015.

153 P. D. Bianco, P. Ducheyne and J. M. Cuckler, *J. Mater. Sci.; Mater. Med.*, 1997, **8**, 525.

154 A. B. Ferguson Jr., P. G. Laing and E. S. Hodge, *J. Bone Jt. Surg.*, 1960, **42A**, 76.

155 J. L. Woodman, J. J. Jacobs, J. O. Galante and R. M. Urban, *J. Ortho. Res.*, 1984, **1**, 421.

156 J. A. Hunt, D. F. Williams, A. Ungersbock and S. Perrin, *J. Mater. Sci.; Mater. Med.*, 1994, **5**, 381.

157 P. A. Lilley, D. R. May, P. S. Walker and G. W. Blunn, in *Advances in Biomaterials*, Vol. 10, ed. P. J. Doherty, Elsevier, Amsterdam, 1992, p. 153.

158 R. L. Buly, M. H. Huo, E. Salvati, W. Brien and M. Bansal, *J. Arthro.*, 1992, **7**, 315.

159 C. A. Bain and P. K. May, *Int. J. Oral Maxillofac. Implants*, 1993, **8**, 609.

160 P. J. Brockhurst and V. McLaverty, *Aust. Dent. J.*, 1981, **26**, 287.

161 F. E. Shepard, G. C. Grant, P. C. Moon and L. D. Fretwell, *J. Am. Dent. Assoc.*, 1983, **106**, 198.

162 B. Bergman, C. Bessing, G. Ericson, P. Lundquist, H. Nilson and M. Andersson, *Acta Odontol. Scand.*, 1990, **48**, 113.

163 M. Yamauchi, M. Sakai and J. Kawano, *Dent. Mater.*, 1988, **7**, 39.

164 M. Kononen, J. Rintanen, A. Waltimo and P. Kempainen, *J. Prosthet. Dent.*, 1995, **73**, 4.

165 I. Watanabe, J. H. Watkins, H. Nakajima, M. Atsuta and T. Okabe, *J. Dent. Res.*, 1997, **76**, 773.

166 J. Takahashi, H. Kimura, E. P. Lautenschlager, J. H. Chern Lin, J. B. Moser and E. H. Greener, *J. Dent. Res.*, 1990, **69**, 1800.

167 O. Miyakawa, K. Watanabe, S. Okawa, S. Nakano, M. Kobayashi and N. Shiokawa, *Dent. Mater. J.*, 1986, **8**, 175.

168 F. Watari, *J. Jpn. Dent. Mater.*, 1989, **8**, 83.

169 C. Ohkutbo, I. Watanabe, J. P. Ford, H. Nakajima, T. Hosoi and T. Okabe, *Biomaterials*, 2000, **21**, 421.

170 T. Albrektsson, P.-I. Branemark, H.-A. Hansson and J. Linstrom, *Acta Orthop. Scand.*, 1981, **52**, 155.

171 J. Doundoulakis, *J. Prosthet. Dent.*, 1987, **58**, 471.

172 A. P. Ameen, R. D. Short, R. Johns and G. Schwach, *Clin. Oral Impl. Res.*, 1993, **4**, 1444.

173 A. P. Ameen, R. D. Short, C. W. I. Douglas, R. Johns and B. Ballet, *J. Mater. Sci.; Mater. Med.*, 1996, **7**, 195.

174 J. F. Bates and A. G. Knapton, *Int. Metals Rev.*, 1977, **22**, 19.

175 R. G. Craig, *Resorative Dental Materials*, 10th edition, Mosby-Year Book Inc., St Louis, MO, 1997.

176 L. Gettleman, *Curr. Opin. Dent.*, 1991, **2**, 218.

177 B. I. Johansson, J. E. Lemons and S. Q. Hao, *Dent. Mater.*, 1989, **5**, 324.

178 S. Mischler, A. Vogel, H. J. Mathieu and P. Landolt, *Corr. Sci.*, 1991, **32**, 942.

179 H. J. Mueller, J. W. Lenke and M. S. Bapna, *Scan. Microsc.*, 1988, **2**, 777.

180 C. D. Wright, R. M. German and R. F. Gallant, *J. Dent. Res.*, 1981, **60**, 809.

181 L. Reclaru and J.-M. Meyer, *J. Dent.*, 1995, **23**, 301.

182 L. Reclaru and J.-M. Meyer, *J. Dent.*, 1994, **22**, 159.

183 D. J. L. Treacy and R. M. German, *Gold Bull.*, 1984, **17**, 46.

184 M. Pourbaix, *Atlas of Electrochemical Equilibria*, National Association of Corrosion Engineers, Houston, TX, 1966.

185 P. Thomsen, C. Larsson, L. E. Ericson, L. Sennerby, J. Lausmaa and B. Kasemo, *J. Mater. Sci.; Mater. Med.*, 1997, **8**, 653.

186 D. F. Williams, *J. Med. Eng. Technol.*, 1977, **1**, 266.

187 E. H. Greener, *Oper. Dent.*, 1979, **4**, 24.

188 K. Anusavice, *Phillips Science of Dental Materials*, W. B. Saunders Inc., Philadelphia, PA, 1996.

189 N. K. Sarker and J. R. Park, *J. Dent. Res.*, 1988, **67**, 1312.

190 D. Brown, Dental amalgam, in *Dental Materials and Their Clinical Applications*, ed. H. J. Wilson, J. W. McLean and D. Brown, British Dental Journal, London, 1988, Chapter 5.

191 D. B. Mahler, *J. Dent. Res.*, 1997, **76**, 537.

192 D. B. Mahler, J. D. Adey and M. Marek, *J. Dent. Res.*, 1982, **61**, 33.

193 E. N. Gale, J. W. Osborne and P. G. Winchell, *J. Dent. Res.*, 1982, **61**, 678.

194 G. W. Marshall, B. L. Jackson and S. J. Marshall, *J. Am. Dent. Assoc.*, 1980, **100**, 43.

195 T. R. Pitt Ford, *The Restoration of Teeth*, 2nd Edition, Blackwell, Oxford, UK, 1992.

196 P. J. J. Plassmans, N. H. J. Creugers and J. Mulder, *J. Dent. Res.*, 1998, **77**, 453.
197 H. M. Pickard, *Manual of Operative Dentistry*, 5th Edition, Oxford University Press, Oxford, UK, 1983.
198 H. Letzel, M. A. van't Hof, M. M. A. Vrijhoef, G. W. Marshall and S. J. Marshall, *Dent. Mater.*, 1989, **5**, 115.
199 J. G. Bauer and H. A. First, *Calif. Dent. Assoc. J.*, 1982, **10**, 47.
200 G. A. Holland and K. Asgar, *J. Dent. Res.*, 1974, **53**, 1245.
201 N. K. Sarker and E. H. Greener, *J. Oral Rehabil.*, 1975, **2**, 49.
202 M. Merck, *J. Dent. Res.*, 1997, **76**, 1308.
203 L. Björkman, N. Pedersen and P. Lichtenstein, *Comm. Dent. Oral Epidemiol.*, 1996, **24**, 260.
204 J. Clausen, *Acta Neurol. Scand.*, 1993, **87**, 461.
205 A. H. B. Schuurs, *J. Dent.*, 1999, **27**, 249.
206 C. O. Enwonwu, *Env. Res.*, 1987, **42**, 257.
207 A. Ericson and B. Kallen, *Int. Arch. Occup. Environ. Health*, 1989, **61**, 329.

CHAPTER 5

Dental Materials

1 Introduction

The destruction of the mineralised structure of the tooth through the metabolic activities of oral micro-organisms is known as dental caries, and has been described in Chapter 1. Active caries, that is, caries in which there is detectable demineralisation, has been shown to have a pH of 4.9 ± 0.2, but this is arrested by increasing the pH to 5.7 ± 0.5.[1] The main acid found to be present in active caries is lactic acid ($88.2 \pm 8.3\%$), with minor amounts of acetic (ethanoic) acid ($9.6 \pm 5.9\%$).[2] Caries in enamel is straightforward, owing to the almost completely inorganic nature of the substrate. In dentine, however, it is more complicated and, in addition to the demineralisation *via* attack at the inorganic phase, there is destruction of the organic components.[2]

The classic work on the relationship between sugar and caries was carried out by Stephan, who reported his results in 1940.[3] More recent studies have given additional insight into the mechanism of the disease, including the role played by minor acids, such as ethanoic and propanoic.[2] It has also been shown that a further feature of dental caries is that the metabolism of sugars causes the oral micro-organisms to form intracellular and extracellular poly-saccharides.[1] These enable the bacteria to cling tenaciously to the surface of the tooth, and also provide a source of energy for continued metabolic activity.

As part of the preparation of the tooth for repair following the identification of active caries, the affected material needs to be removed. In modern clinical practice, this is usually done using a carbon-steel bur driven at high speed by a dental hand-piece.[4] This leads to the development of a smear layer, an amorphous layer of organic and inorganic debris formed on the dentine. It is between 3 and 10 μm thick, adheres strongly, and cannot be removed by the ordinary water spray.[5,6]

Certain dental materials, most notably amalgam restorative materials and titanium implant materials, have already been considered in previous chapters. In the current chapter, the emphasis is on tooth-coloured dental restorative materials. These materials are of widely varying chemical types, and serve a variety of functions within clinical dentistry. These include direct filling materials, but also prosthodontic applications, such as bonding of crowns and bridges, luting and so-called core build-up, *i.e.* the reinforcement of the core part of a tooth, prior to the replacement of a metal or ceramic crown.

The materials to be considered are classifed as follows.

Composite resins: These are two-phase materials consisting of polymeric continuous phase (matrix) and powdered inorganic filler, typically silica of some sort. Setting occurs by polymerisation, either chemical or more usually light-initiated, and there is no chemical reaction between the filler and the matrix.

Cements: These are also two-phase materials, where the continuous phase is formed by reaction of an aqueous solution of acid, either phosphoric or a water-soluble polymer, such as poly(acrylic acid), with a powdered base, either zinc oxide or a special basic glass.[7] The matrix is thus a salt, and the set cement derives some of its mechanical properties from the presence of unreacted base powder, which acts as reinforcing filler. Examples include zinc phosphate, zinc carboxylate and glass-ionomers. The most versatile of these are the glass-ionomers, and their use includes as direct filling materials in their own right. In addition, uniquely among cements, they have been modified by the inclusion of monomers, so that they can be cured by a dual process of neutralisation and addition polymerisation.

Polyacid-modified composite resins: Informally known as *compomers*, these are hybrid materials that are predominantly of the composite resin type.[8] They consist of a polymerisable matrix and inert filler, but also include special monomers bearing acid groups, and also glass powders of the glass-ionomer type. Their setting is entirely by polymerisation, but later they take up small amounts of water, which initiates the acid–base reaction, and promotes some salt formation.

As research continues on modern materials, so the boundary between the extremes, the composite resins and widely used restorative cement, the glass-ionomer, is blurring to some extent.[2,9] However, there is a critical distinction between the two fundamental groups of materials, namely their water content. Those materials which are water-based, *i.e.* glass-ionomers and their resin-modified counterparts, are significantly more hydrophilic than those materials based on organic systems. This has the clinical consequence that they wet the high energy surface of the freshly cut tooth very readily and, over time, develop a strong natural bond with the vital tooth. The composite resin materials, including the polyacid-modified versions, by contrast, are hydrophobic. They therefore do not wet the tooth surface, and require more elaborate preparation procedures in order to develop a durable bond to the tooth surface.

These materials are now considered individually in detail in the rest of this chapter.

2 Composite Resins

Although the term *composite* generally refers to any material consisting of two or more phases, within dentistry it has a more restricted use. It refers to a

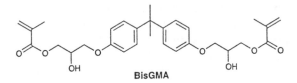

Figure 5.1 *Structure of bisGMA*

material which sets from a paste to a hardened material by a polymerisation mechanism, without involving any chemical reaction between the filler and the matrix. These materials are thus considered distinct from cements, which harden by reaction of the filler, which is basic, with the an acidic liquid to form a salt matrix. Although the set cement includes unreacted base within it as reinforcing filler, and so is properly a composite material, it is not usually described as such within the terminology of dental materials.

Modern dental composite restorations can be considered to have originated in Bowen's classic work on the development of the monomer 2,2-bis-4-(2-hydroxy-3-methacryloyloxypropoxy)phenylpropane.[10] The essential chemistry of the synthesis of this substance is the reaction of glycidyl methacrylate with bis-phenol A to create a molecule known informally as bis-glycidyl methacrylate or bisGMA (Figure 5.1).

The original aim of the synthetic studies of Bowen that led to the development of this monomer was to combine the advantages of the acrylic systems with those of epoxy systems based on bisphenol A but without the disadvantages. This was largely successful, and bisGMA is the basis of most current composite resin systems used clinically.

A difficulty with bisGMA is that it is very viscous at room temperature, a result of the presence of OH groups in the molecule,[11] and this makes handling and incorporation of fillers difficult. Consequently it is usually diluted with lower molecular weight, lower viscosity monomers, such as triethylene glycol dimethacrylate, TEGDMA. The reduction in viscosity achieved with these diluents is remarkable, as can be seen in Table 5.1. The disadvantage of adding diluent is that it tends to increase the contraction on polymerisation. This effectively limits the amount of diluent that can be added to a practical formulation for clinical use.

The problem of polymerisation contraction is an inherent feature of polymer-based restorative systems. However, it can be significantly reduced if not overcome completely by the inclusion of fillers in the material, the chemistry of which is discussed later in this chapter.

Curing of Composites

Two mechanisms are available for curing composite resins, and these are generally distinguished as *chemical cure* and *light-activation*.

Chemical cure types are two-paste systems, which require mixing prior to

Table 5.1 *Viscosities of bisGMA/TEGDMA blends*[12]

Blend (BisGMA : TEGDMA)	Viscosity/cP
75 : 25	4300
50 : 50	200

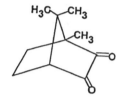

Figure 5.2 *Structure of camphorquinone*

placement. Mixing is a critical step, as it must be sufficiently thorough that the activator (*N,N*-dimethyl-*p*-toluidine) in one paste comes into contact with the free-radical initiator (benzoyl peroxide) in the other. On the other hand, this process can incorporate air into the material, thus creating voids that reduce the strength of the set composite. This is now more or less obsolete, at least for restorative grade composite resins, and most current materials make use of light-activation.

Light-activated composites are one-paste systems, which contain a photo-initiator that is sensitive to blue light (*i.e.* light of wavelength 470 nm). Setting is initiated by the use of a special dental curing lamp that emits light at this wavelength and with a light intensity[13] of at least 500 W m^{-3}. However, care is needed with these materials, as they will undergo some sluggish polymerisation under normal surgery lights.[14]

The initiator used in these materials is camphorquinone (bornanedione, 1,7,7-trimethylbicyclo[2.2.1]heptane-2,3-dione), typically in association with amines.[15] Its structure is shown in Figure 5.2.

Camphorquinone absorbs light in the region 200–300 nm and at 470 nm, the latter being responsible for its yellow colour. Both absorptions occur in the dicarbonyl group, the one at 200–300 nm as a result of a freely allowed π,π^* transition, the one at 470 nm as a result of a symmetry-forbidden n,π^* transition. The different natures of the two transitions are reflected in the ε_{max} values of 10 000 and 40 respectively.[16]

The photochemistry of camphorquinone is complicated. In oxygen-free atmospheres, it is reduced to α-hydroxyketones,[17,18] whereas in the presence of oxygen it gives rise to a variety of oxidation products, including esters of camphoric acid, camphoric anhydride, acids and α-hydroxyketones.[19]

When camphorquinone is irradiated in the presence of hydrogen donors, such as the amine accelerators used in dental composites, there is an initial abstraction of hydrogen by the excited camphorquinone molecule, CQ*. The radicals thus formed from the hydrogen donors can initiate the required polymerisation with the bisGMA and other monomers. A typical accelerator is ethyl 4-dimethylaminobenzoate, EDAB. Studies have shown[17] that there is a rapid reaction:

$$EDAB + CQ^* \rightarrow EDAB(-H)\cdot + CQH\cdot$$

In the presence of oxygen this is quickly followed by:

$$EDAB(-H)\cdot + O_2 \rightarrow \text{peroxidation chain}$$

Similar reactions are able to occur with some of the monomers used, such as TEGDMA, though the hydrogen-donating ability of these molecules is less than that of amine co-initiators. The presence of oxygen has been shown to exert a positive effect on polymerisation processes initiated by camphorquinone. This is because peroxy groups may be formed and these can undergo photodissociation assisted by the camphorquinone.[20] This reaction generates new radicals which may become involved in the initiation process, causing it to be accelerated.[21] This is in contrast with what is observed with other initiators, where the presence of oxygen inhibits the polymerisation of acrylic monomers.[22]

In clinical use, care has to be taken over depth of cure with light-activated composites. Light is attenuated as it passes through the composite, which means that there is less light to activate polymerisation deep within the material. This in turn means that the upper layer cures better than the lower ones,[23] and in the extreme, the very bottom may not cure at all. In addition, the colour of the restoration affects the depth of cure, since darker shades attenuate light more readily than lighter ones. To overcome these problems, the technique of incremental build-up of restorations is used, each increment being 2 mm or less, to ensure there is adequate light-penetration even at the bottom of the layer.

The degree of conversion of bisGMA monomer under clinical conditions has been shown to be approximately 55% for a chemically cured system[24] and about 48% for a UV cured system.[25] Use of diluent can increase this to more than 70%. This increase in conversion leads to improvements in the stability of the cured resin and to an increase in the degree of crosslinking, which in turn brings about an increase in modulus, due to reduced mobility of polymer chains.

Fillers

Fillers, based on silicas or silicates of various types are blended into the monomer phase to produce the composite system.[26] Fillers confer many desirable properties. They reduce the overall contraction due to polymerisation, and also reduce water uptake and coefficient of thermal expansion. They improve mechanical properties of the cured material (compressive strength, tensile strength, Young's modulus and abrasion resistance).

Table 5.2 *Classification and properties of dental composite restorative materials*[29]

Material: Property	Traditional (macrofilled)	Microfilled	Small particle	Hybrid
Particle size/μm	8–12	1–5	0.04–0.4	0.6–1.0
Filler vol. (%)	60–65	20–55	65–75	60–65
Filler wt. (%)	70–80	35–60	80–90	75–80
Compressive strength/MPa	250–300	250–350	350–400	300–350
Tensile strength/MPa	50–65	30–50	75–90	70–90
Young's modulus/GPa	8–15	3–6	15–20	7–12
Hardness (Knoop no.)	55	5–30	50–60	50–60

Quartz has been used as the material for filling these composite systems, though more recently it has been replaced by improved fillers fabricated from strontium and barium glasses. Their advantages include that they are softer, can be more readily prepared at lower particle sizes, and they possess the property of radiopacity. There have been concerns that barium glasses may prove toxic,[27] although this does not seem to have been a problem to date. On the other hand, their hydrolytic instability in the moist environment of the mouth has proved a problem, leading to increased leaching of barium ions and increased rate of degradation of the composite.[28]

The particle size of these fillers varies widely, from of the order of 0.04–100 μm. The larger size fillers are prepared by grinding or milling quartz or glasses, and the resulting powders have particles in the range 0.1–100 μm. Colloidal silica particles (*i.e.* those in the 0.04 μm size range) may be prepared by pyrolytic processes in which $SiCl_4$ is burned in an atmosphere of oxygen and hydrogen. Alternatively, they can be prepared from colloidal SiO_2 or by reacting colloidal sodium silicate with HCl. The resulting fine dust-like particles of silica are spheroidal in shape, and are referred to as a microfiller. Their very small size means there is a severe limitation on the amount that can be incorporated into the monomer blend without producing an excessively viscous paste.

The filler particle size and distribution are important features since, by manipulating them, the properties of composite materials can be varied widely. Briefly, the durability, especially wear, is improved by having very fine particle size fillers. However, the volume of such filler that can be incorporated is much lower than for the larger particle size fillers, hence the composites containing these fillers tend to have reduced filler volume fractions and this in turn leads to lower strengths. The classification of dental composites is shown in Table 5.2, together with typical properties.

The wear characteristics of traditional composites tended to be poor, and this is the reason that they have been superseded by composites containing more complex arrays of particle sizes. It was found that, on polishing, or in normal service (tooth-brushing and mastication), the softer organic matrix became worn away, leaving large abrasive filler particles protruding from the surface, giving it a rough morphology. This roughened surface was not only unattractive from the aesthetic point of view, it was also susceptible to staining. Small particle and hybrid composites, by contrast, have much improved wear characteristics, and can be used in situations where good resistance to wear is desirable. Appearance remains acceptable because of the improved wear, and also these materials are less prone to staining than traditional composites.

Fillers are required to be properly bound into the matrix if they are to provide mechanical reinforcement. This enables the stress to be transferred from the lower modulus polymeric phase to the higher modulus filler, and is achieved by using a coupling agent.[29] The most extensively used of these is γ-methacryloxypropyltrimethoxysilane:[30]

$$CH_2=C(CH_3)CO.OCH_2CH_2CH_2Si(OCH_3)_3$$

This substance is deposited onto the surface of the filler using a 0.025–2% aqueous solution which is sufficient to create a monolayer. Adhesion is achieved by the siloxane bond between the silanol groups of the hydrolysed silane and the equivalent functional groups on the surface of the filler. The silane triols become adsorbed on the surface of the filler as monomeric or oligomeric layers as a result of the condensation reaction between the coupling agent and the surface groups. It is not only adhesion of filler to matrix that is enhanced by the presence of these coupling agents but hydrolytic degradation of the interface has also been shown to be reduced when these silanes are present.[32]

Pigments

The usual pigments for matching the shade of the tooth are iron oxides.[26] They give good colour, which is stable for long periods in the hostile oral environment, and they are sufficiently varied that a wide range of shades can be prepared. In addition, optical brighteners which fluoresce under weak UV light are added to give a more lifelike appearance, especially in artificial lighting, which would otherwise give the restoration a dull, unaesthetic appearance.

Bonding of Composite Resins to the Tooth

Bonding of these materials varies according to the substrate. For enamel, the acid-etch technique has been used for many years, and still is by many practitioners.[33] In this technique, enamel is treated with a phosphoric acid gel at 37% concentration for 20–30 seconds, after which the surface is cleaned and dried. The drying reveals a frosted appearance, which arises from exposure of enamel prisms, and this surface has an increased surface roughness. Onto this surface is placed a layer of unfilled blended bisGMA/TEGMA, which flows into the surface irregularities and polymerises to give an essentially hydrophobic surface that is kept in place by so-called micro-mechanical retention.

Resulting bond strengths are good, typically in the region 20–24 MPa or higher for shear bonding to human enamel.[34]

For dentine, the bonding process is much more complicated.[35] This is because the dentine is closer to the living centre of the tooth, the pulp, than the enamel and has running through it numerous tubules which connect with the pulp. These tubules are full of fluid, and there is a tendency for this fluid to flow out of the tubules when the tooth is cut. Bonding is thus required to develop under much wetter conditions than those involving just the enamel of the tooth.

When the tooth is cut, a layer of cutting debris develops at the surface. This is known as the 'smear layer', and opinion is divided as to whether or not it should be removed. Certain bonding agents are claimed to be able to make use of this source of calcium, and hence are recommended to be used over the top of the smear layer.[36] Against this, there is considerable experimental evidence that bond strengths to dentine are stronger where the smear layer has been removed.[37]

The method by which the smear layer is removed is also subject to controversy among clinicians. Because of the sensitivity of the nearby pulp, many authorities have recommended that the smear layer be removed with a mild agent, such as citric acid or EDTA. By contrast, others have shown that, provided the dentine layer is intact and very short exposure times are used, etching with phosphoric acid gel is acceptable.[38,39] The use of phosphoric acid for the preparation of both enamel and dentine surfaces is known as *total etch* and is now considered the preferred technique for bonding to the tooth under clinical conditions. Even when milder acids have been used, cleaning away of the smear layer involves not only cleaning but also some modest etching of the surface, and one mode of action of the bonding agent is to flow into the enhanced surface area, as well as into the dentinal tubules[40] which are exposed by the cleaning process. There they harden to form tags which contribute to the retention of the bonding agent.[41]

In addition to tag-formation, there is the formation of resin-dentine hybrid zone,[42] a structure some 2–5 μm thick.[43] This is formed by the infiltration of unpolymerised bonding agent into the collagen fibrils following demineralisation of the intertubular and peritubular dentine by the acid,[44] followed by its subsequent hardening. The development of this hybrid layer between the dentine and the bonding agent[45] has been shown to be critical to the promotion of a durable bond between the tooth and the composite resin,[46–48] Experimental bonding agents have been prepared with a viscosity too great to allow them to flow into the dentinal tubules to form tags. However, they are able to form a detectable hybrid layer and they have been shown to give strong and durable bonds between the composite resin and the tooth.[49] In order to ensure this hybrid layer forms properly, the demineralised collagen must not be dried too thoroughly, otherwise the collagen fibrils collapse, and the liquid monomer is not able to penetrate. This leads to inferior bond strengths.[50]

There are currently a wide variety of bonding systems available to the clinician. They share the characteristic of being formulated from molecules

Table 5.3 *Effect of saliva contamination on shear bond strength*

Condition	Strength on enamel/MPa	Strength on dentine/MPa
Clean	31.7 ± 4.2	18.8 ± 3.2
Contaminated	17.9 ± 5.6	3.8 ± 2.2

having both polar and non-polar functional groups. One substance used extensively in these formulations is 2-hydroxyethyl methacrylate, HEMA, but other molecules have been used, including a range of organophosphorus compounds.[51] The most successful systems have been those based on multiple components and multiple applications,[52] so that the dentine surface has been etched, primed and then sealed prior to bonding. Now, as a result of the demand from clinicians for faster treatments, so-called 'single-bottle' bonding agents are becoming available designed to combine all the functions of the individual components in a single formulation,[53] one preferably applied only once. The danger with this approach is that by simplifying the procedure, the bond to dentine may be compromised,[54] either in strength or durability.

Clinical studies of durability of composite resins indicate that these procedures are satisfactory and that there are few if any adhesion failures. Surface cleanliness of the tooth is vital, and inadvertent contamination of the freshly cut surface with saliva can radically reduce the strength of the bond that is subsequently formed with that surface.[55] Data to illustrate this point are shown in Table 5.3, where surfaces of dentine and enamel were prepared, treated with three layers on bonding agent, and dried before a commercial composite resin was applied. One set was deliberately contaminated by exposure to saliva for 15 seconds, followed by drying for 5 seconds, while the other set was bonded using the ideal clinical procedure. Specimens bonded to contaminated surfaces were found to have significantly lower bond strengths when they were subsequently tested in shear and also to change their failure mode from cohesive to interfacial.

Contamination of the freshly cut tooth surface by blood, saliva or even oil from the air compressor will alter the surface energy. This in turn affects the ability of the bonding agent to adhere to the surface.[56] Clinicians are therefore advised that, in the event of accidental contamination of freshly cut surfaces, the tooth should be re-etched, though for a shorter time than for initial treatment, *e.g.* for 10 seconds rather than up to 60 seconds.

The fact that composite resins undergo contraction on polymerisation creates the possibility of a gap forming at the margin of the cavity and the restoration. This allows the possibility that bacteria and oral fluids may penetrate, causing damage to the pulp and possible inflammation.[57] This can be very painful for the patient. The phenomenon has been termed *marginal leakage*, a term which implies that the margins suffer from relatively gross imperfections. With improved bonding systems, and the lack of visible gaps, it might have been assumed that the problem was solved, but in fact microscopic gaps often remained, and the resulting leakage was termed *microleakage.*[58]

This now seems to have been eliminated with the use of the technique of total-etch, together with the latest generation of bonding agents. However, there remains a slight permeability in the bond to dentine, and this is referred to as *nanoleakage*.[59] This does not seem to cause clinical problems, possibly because bacteria cannot penetrate, though it has been detected in a number of bonding systems.[60]

Clinical Performance of Composite Resins

Composite resins have been successful in a variety of clinical situations, as already indicated, and they show reasonable longevity in these situations. However, their wear characteristics are inferior to those of amalgams. Clinical wear is a complex phenomenon, and involves abrasion, fatigue and chemical attack, such as hydrolysis and erosion.[61]

In composites, these phenomena act together to lead to progressive exposure of the axially directed cavity wall. Wear behaviour varies with the type of composite. Traditional composites usually wear most at the margins, whereas microfilled composites exhibit chipping and microfracture across their entire surface.

Composite resin systems are used for applications other than direct filling of the tooth. These other applications are:

(a) Luting: The word lute comes from the Latin for mud (*lutum*), and suggests that the material is present simply to fill space and prevent the entrance of fluids; no adhesion is implied. Resin-based luting cements are based on very similar chemistry to that of composite resins in that they consist of a resin, predominatly bisGMA or urethane dimethacrylate, plus a filler of finely divided silica. However, the filler loading is much lower, which gives them a lower viscosity. They have high tensile strength, and this makes them suitable for micro-mechanically retaining etched ceramic veneers, and also for fixed partial denture retainers to tooth preparations that would otherwise be insufficient for retention using conventional cements.

Luting composites come in a variety of types. They may be chemically-cured, for use under opaque restorations, such as metal crowns, or they may be light-activated for use under translucent ceramic veneers. Care has to be taken that very thin layers of luting composite are used; otherwise setting shrinkage can lead to severe marginal leakage.[62]

(b) Bonding of bridges: As with the bonding of composite resin restorations of individual teeth, the fixation of bridges relies on the technique of acid-etching. The first resin-bonded appliances were described by Rochette.[63] They were splints held in place by an unfilled acrylic resin attached to etched enamel. Subsequently, other devices, such as the Maryland Bridge, have been developed, and these may be bonded in place with an unfilled or lightly filled composite resin based on bisGMA.[64,65] The properties and methods of use of the commercial resin bonding materials vary somewhat, and clinicians are advised that careful attention should be paid to the manufacturers' instructions. It is also essential to remove any excess composite from around the

Table 5.4 *Components of dental cements*

Cement	Acid	Base	Clinical uses
Zinc phosphate	Phosphoric	Zinc oxide	Luting; bases
Zinc oxide–eugenol	Eugenol	Zinc oxide	Temporary restorations; pulp capping; root canal therapy
Zinc polycarboxylate	Poly(acrylic)	Zinc oxide	Luting; bases
Glass-ionomer (GI)	Poly(acrylic)	Glass	Luting; liners and bases; anterior restorations

edges of the appliances since, once hardened, such excess material may cause irritation to the gingival tissue.

(c) Core build-up: Quite frequently, prior to placing a crown, dentists are faced with the need to build-up a core structure of a tooth that has been extensively damaged by trauma or by caries. This can be done using a composite resin, and there are several advantages in using composites rather than amalgam, as has been traditional. For example, though they have fracture toughness properties similar to amalgam, they are easier to place, and (unlike amalgam) allow further preparation of the tooth to be continued immediately after placement.

3 Cements

Basic Types and Uses

Dental cements are all based on mixtures of acids and bases, which set by a process of neutralisation to give a salt matrix and leave some unreacted base which acts as reinforcing filler. Details of the commonly used cements appear in Table 5.4.

Cements Based on Zinc Oxide

Dental cements use a special form of zinc oxide that has been deactivated by mixing with magnesium oxide (approximately 10%) followed by being sintered together at temperatures between 1000 and 1400 °C. This sintering produces a cake which is then reground to a fine powder for use in the cement. The process of heating causes the zinc oxide to become slightly yellow in colour and less reactive. This is due to evaporation of oxygen and the formation of a slightly non-stoichiometric oxide corresponding to $Zn_{(1+x)}O$, where x is up to 70 ppm.[66]

(a) Zinc phosphate. The so-called zinc phosphate cement has a long history of use in dentistry, having appeared before the end of the 19th century. The liquid for this cement is based on aqueous phosphoric acid, typically in the

concentration range 45–65%, containing aluminium (1.0–3.1%) and possibly zinc (up to 10%).[67] These metal additives have the effect of moderating the rate of reaction between the acid and the base, and they also reduce the heat of reaction. The properties of the set cement, both mechanical and chemical, depend critically on the concentration of H_3PO_4 in the liquid.[68] For this reason, the liquid component must be stored carefully, and not allowed to gain or lose moisture to the atmosphere.

For zinc phosphate, there have been numerous previous studies of the setting process. It is known that there is an initial acid–base reaction that gives rise to a soluble acidic zinc phosphate:[69]

$$ZnO + 2H_3PO_4 \rightarrow Zn(H_2PO_4)_2 + H_2O$$

Further reaction of this acidic zinc phosphate leads to the rapid formation of the insoluble zinc orthophosphate, $Zn_3(PO_4)_2$. This latter substance is essentially amorphous, and is prevented from crystallising by the presence of aluminium in the cement-forming liquid.[70] However, over long periods of time, some of it does crystallise, forming the hydrated phase $Zn(H_2PO_4)_2.4H_2O$ known as hopeite.[71] Growth of hopeite, which occurs at the surface, is favoured by elevated temperatures and high humidity, though even after over forty years in the mouth, this crystalline component has been shown to represent only a small fraction of the cement mass.[12]

As the zinc phosphate cement sets, there are considerable changes in the amount of soluble material that may be eluted.[10] Relatively little zinc dissolves at any time, the major soluble species being phosphate. The amount of this phosphate depends *inter alia* on the pH of the storage solution, which itself depends on the amount of material leached. Solubility has been shown to drop rapidly as pH rises from 4.5 to 5.0, then to rise again more slowly as neutral conditions are approached.[72]

(b) Zinc oxide–eugenol. Zinc oxide–eugenol is used in dentistry for the temporary cementation of crowns, as a root canal sealant, as a sedative cavity liner and in soft tissue packs in oral surgery.[73] It has a long history of use in dentistry, dating back to at least 1875.[74] Eugenol is a weak acid with a pK of 10.4,[75] and occurs as hydrogen-bonded dimers.[76] Cement formation is the result of an acid–base reaction in which traces of water play an essential role, and which results in the formation of a zinc eugenolate chelate. Accelerators, such as 0.1–2% acetic acid, are added to the eugenol to ensure a sufficiently rapid reaction for clinical use.[77] The set cement has relatively poor mechanical strength, *i.e.* compressive strength of 13–38 MPa, but this is sufficient for its use for temporary cementation.[78] Zinc oxide–eugenol also finds some use as a dental impression material, mainly for correcting impressions taken in other materials for which a detail is not clear.

(c) Zinc polycarboxylate. The zinc polycarboxylate cement consists of a similar modified zinc oxide powder to that used in the zinc phosphate cement.

Table 5.5 *Properties of the zinc polycarboxylate dental cement*

Property	Typical component/values
Liquid	40–50% polyacid
Polymer	Usually poly(acrylic acid)
Powder:liquid ratio	2.5–3 : 1
Setting time/min	2.5–4
Compressive strength/MPa	80–100

It differs from the zinc phosphate cement in that it employs a water-soluble polymeric acid in place of phosphoric acid, a variation which makes the resulting cement adhere to dentine and enamel. Indeed, the development of the zinc polycarboxylate cement was an important one in the history of adhesive dental materials.

Zinc polycarboxylate cements are used for a variety of procedures within clinical dentistry, including the lining of cavities, attaching crowns to posts, and bonding orthodontic brackets in place. Typical properties of the zinc polycarboxylate cement are shown in Table 5.5.

The structure of the set cement has been studied in detail. Early indications, using infrared spectroscopy, were that this material was purely ionic, with Zn^{2+} ions showing no interaction with the adjacent carboxylate groups.[79] However, this proved to be an artefact of the poor resolution of technique, because when this material was studied using FTIR a complex array of structural types were detected, all showing some partial covalency in the interaction of Zn^{2+} ions with carboxylate groups.[80] The various structures proposed are illustrated in Figure 5.3.

The mechanical properties of the zinc polycarboxylate cement are different from other cements of this class, in that the material has some distinctly plastic character.[81] Indeed, their behaviour can be modelled as that of a thermoplastic composite, which implies that there are only weak crosslinks between the polymer molecules.[82]

The role of water in these materials has been studied in some detail. Water is involved in the setting reaction, because it is needed to dissolve the poly(acrylic acid). Since setting takes place without phase separation, it is known that water remains within the set cement, at least under normal circumstances. Storing these cements under desiccating conditions, either dry air[83] or in saturated salt solutions[84] causes these cements to lose considerable amounts of mass, presumably due to loss of water, but this has not been found to affect their compressive strength. This led to the conclusion that water, which probably occupies co-ordination sites around the zinc ions and the polyanions, does not have a structural role.

An important property of zinc polycarboxylate cements is their adhesion to dentine and enamel, which enables them to be used as dental adhesives. However, the literature values for bond strength vary widely. This was explained by Akinmade and Hill, who suggested that the probable source of

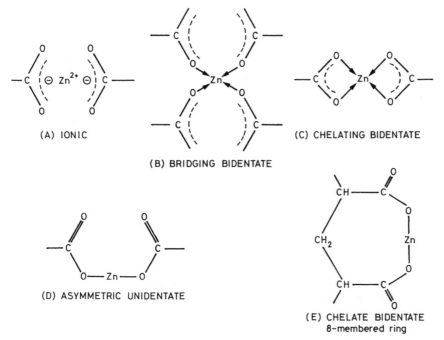

Figure 5.3 *Possible chelate structures occurring in zinc polycarboxylate dental cements*

scatter was the failure to control the thickness of the adhesive layer.[85] They studied the effect of this on the measured bond strengths of zinc polycarboxylate cement, using graded glass ballotini mixed into the cement to control the thickness of the film. They found that there was an optimum thickness corresponding to a plastic zone of comparable size. This plastic zone varied with molecular weight of the polymer, from 40 μm at M_w of 1.15×10^4 to 290.9 μm at M_w of 3.83×10^5. The existence of an optimum adhesive layer thickness was confirmed for a commercial zinc polycarboxylate cement, using a more sophisticated arrangement of external spacers to control the thickness[86] (Figure 5.4). This technique had the advantage of not adding foreign materials to the cement, though it also served to confirm the validity of the original observations.

These results showed that the commercial cement had a plastic zone size comparable to those in the experimental cements derived from poly(acrylic acid) samples of similar molecular weight. It also demonstrated that thicker layers were more likely to fail cohesively than thinner ones.

4 Glass-ionomers

These are probably the most versatile of all the dental cements, as shown in Table 5.4. They may be considered to be related to the zinc polycraboxylate in that they employ a water-soluble polymeric acid, usually poly(acrylic acid),[87]

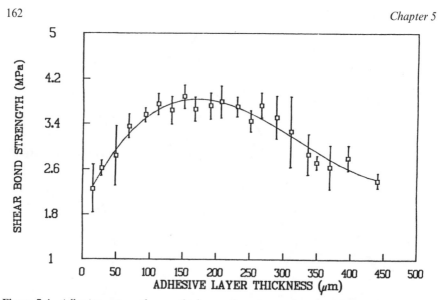

Figure 5.4 *Adhesion strength vs. thickness for zinc polycarboxylate using spacers (optimum = 205 μm)*[86]

as their acidic component. Other polymers, such as copolymers of acrylic and maleic acid[88] and homopolymers of vinyl phosphonic acid have been used in commercial cements,[89] and experimental cements have been made with copolymers of acrylic and itaconic acid.[90] Sulfonic acid polymers, though capable of forming cements, do not yield materials of sufficient hydrolytic stability to be used in dentistry. In addition, tartaric acid is often added to sharpen the setting reaction and improve the overall handling of the cement.[91]

Polyelectrolytes are linear polymers that contain a multiplicity of ionisable functional groups.[92] In aqueous solution, they dissociate into polyions and small ions of opposite charge, known as counterions. As has been mentioned already, the polyelectrolytes used in dentistry are all anionic.

Aqueous solutions of polyelectrolytes behave differently from either solutions of uncharged polymer in organic solvents or from those of low molecular weight solutes in water. This is because of the effects arising from the interactions of electrical charges.[78] These have generally been studied under very dilute conditions and high added salt concentrations, and thus are of little value in understanding the behaviour of polyelectrolyte solutions used in dental cements, where polymer concentrations typically lie in the range 40–50% and there are usually no added salts. However, it is possible to extrapolate to an extent from regions of low concentration and to understand, qualitatively at least, some of the features of practical polyelectrolyte solutions.

Anionic polyelectrolytes tend to adopt helical conformations in aqueous solution,[93] since these minimise electrostatic repulsion. Progressive neutralisation results in a gradual increase in charge density, and this leads to progressive coil expansion. In dilute solutions, this coil expansion continues until the molecule ceases to be helical but has become rod-like. The helix-to-rod

Table 5.6 *Effect of progressive neutralisation of 45% aqueous poly(acrylic acid) with NaOH*[94]

Neutralisation (%)	pH	Density/g cm^{-3}	Viscosity/cP
0	1.38	1.15	100
10	3.16	1.17	409
25	3.83	1.19	907
40	4.33	1.24	2210
55	4.71	1.27	4334
70	5.30	1.33	5330
85	5.84	1.35	13097
100	7.83	1.36	51552

transition usually appears at well defined degrees of neutralisation and can be detected by sharp changes in solution viscosity.[78] In concentrated solutions, the situation becomes complicated by the existence of excluded volume effects. However, there is a considerable effect of neutralisation of concentrated poly(acrylic acid), as shown in Table 5.6. The increases in viscosity and also density are consistent with coil expansion and increased chain entanglement, though any such coil expansion is probably less than can occur in dilute solution, and is unlikely to be due to the complete adoption of rod-like morphology.

Neutralisation within polyelectrolyte dental cements is accompanied by ion-binding.[95] In other words, the counterion becomes strongly associated with the polyanion chain, and this has the effect of reducing the local charge density in the region of the polymer molecule. It thus operates in the opposite direction to the effect of simple neutralisation of the polyelectrolyte. The possible sites that can be occupied by counterions around a polyion have been identified by Oosawa[96] as follows:

(i) Atmospheric, *i.e.* relatively mobile, and occupying regions within the electrostatic field of the polyanion;

(ii) Site-bound, *i.e.* essentially immobile, and closely associated with specific functional groups.

Ion binding has been demonstrated by a number of techniques, including titration,[97] dilatometry,[98,99] and NMR spectroscopy.[100,101] Infrared spectroscopy has been especially useful in determining not only whether specific ions are site-bound, but also the details of the bound structures. For dental polymers, these have been found to vary in quite subtle ways depending on the details of the polymer composition. For example, calcium was found to become site-bound, with chelate bidentate geometry, with acrylic/maleic acid copolymer, but to remain in atmospheric regions with the homopolymer of acrylic acid.[102] Aluminium, by contrast, was found to be site-bound with bridging bidentate geometry with the homopolymer, but to be in atmospheric regions around the copolymer.

Ion binding of this type has been shown to influence the state of water molecules within the complexes.[103] Refractive index measurements have been used to show that, at low degrees of neutralisation, the average distance between ionised functional groups is relatively large, and that rearrangement of water molecules is due solely to the charge on the individual functional group. At each charged site, individual hydration spheres of oriented water molecules are formed, so-called *intrinsic water*. In the case of poly(acrylic acid) at 30% neutralisation, these spheres have a radius of 0.31 nm.[104] Above this degree of neutralisation, the distance between the ionised functional groups falls and the individual hydration spheres begin to coalesce. In the end, they form a single sheath of water molecules along the polyanion. At full neutralisation, this sheath consists of two layers, one with a diameter of 0.5–0.7 nm, the other with a diameter of 0.9–1.3 nm.[90]

Consideration of these fundamental aspects of polyelectrolyte behaviour has been important in understanding the chemistry of setting of glass-ionomers. First, though, it is necessary to describe briefly the composition and nature of the glasses used in these materials.

The glasses used in commercial glass-ionomer cements are all of the aluminosilicate type,[105] and contain additionally calcium, fluoride and usually phosphate ions as well. They are made by melting the required mixture of inorganic ingredients at elevated temperatures (1200–1550 °C). The resulting glass melt is shock-cooled, either by pouring onto a cold metal plate and then into water, or by pouring straight into water. The initial product is a coarse glass frit, and this is ground further, usually in a ball mill under dry conditions, to prepare a fine powder. The particle size requirements are that the glass should be below 45 μm for a filling grade cement and below 15 μm for a fine-grained luting cement, though current commercial materials are often finer than these limits. Further treatment may be necessary to reduce the reactivity of the glass powder, either annealing at 400–600 °C or washing with dilute acid, typically acetic acid at 5% concentration.

A large number of glasses have been studied as cement formers, the main ones being based on the systems $SiO_2–Al_2O_3–CaO$ or $SiO_2–Al_2O_3–CaF_2$.[106] However, other types, for example aluminoborate[107] and zinc silicates[108] have been studied and results reported, and certain commercial cements are known to be formulated with glasses in which calcium has been substituted by strontium. This latter substitution makes no practical difference to cement setting, as the interactions of calcium and strontium with poly(acrylic acid) are similar,[109] and setting and handling characteristics are therefore not affected.

The properties of glasses can be generally understood in terms of the Random Network model.[110] This model works on the basis that the glass consists of a random assembly of individual $[SiO_4]$ tetrahedra linked at the corners to form chains. In the simplest case, these chains would be the only constituents and the resulting glass would be inert, due to its electroneutrality. However, if a proportion of the silicon atoms are replaced by aluminium, the resulting tetrahedra become formally charge-deficient, and it becomes necessary to add some additional cations, notably Na^+ and Ca^{2+}, to balance the charges. Glasses of this structure are then susceptible to acid attack, and hence candidates for practical glasses in cements.

The glasses used in practical cements are often phase-separated. This introduces an important variable into the glass-making process, since the extent to which the phases may seperate depends on the speed of quenching. This step affects the reactivity of the glass, since the phase separation process has the effect of altering the availability of more and less reactive phases, and may result in highly reactive phases being locked away inside the glass, away from the reach of the aqueous polyacid. Phases may consist of discrete droplets, or of two connected phases, one of which forms a connected droplet-like structure within the other, more continuous phase. This is the result of so-called spinodal decomposition.[111] Phase-separated glasses tend to give stronger cements than single-phase cements, though recent studies of mono-phase glasses has shown that it is possible to obtain compressive strengths in excess of 150 MPa even without using phase-separated glasses.[112]

A number of factors are known to influence the final compressive strength and handling properties of glass-ionomer cements. These include the molecular weight of the polymer,[113] the concentration of the acid in aqueous solution,[114] the powder:liquid ratio[115] and the presence of chelating agents,[116] of which (+)-tartaric acid is the only one to have been at all widely used in commercial cements. It is also known that cements undergo maturation reactions, and these often lead to increases in compressive[117] and flexural strength,[118] as well as to an increase in the ratio of bound to unbound water. This latter property has been determined by exposing cements to drying conditions, either elevated temperatures (105 °C) or in a desiccator over concentrated sulfuric acid. The amount of water that can be removed by such treatment declines as the cement ages approaching a limiting value after several months. However, even very mature cements retain a proportion of water that can be removed under these conditions. This water has been termed *evaporable* or *unbound*; the remainder has been called *unevaporable* or *bound*. The nature of the 'binding' over time has not been determined, nor is it known whether the water continues to be present as individual water molecules.

The setting chemistry of glass-ionomers has been studied by a variety of techniques, notably infrared spectroscopy,[119] FTIR,[120] ^{13}C NMR spectroscopy,[121] electron probe microanalysis,[122,123] and pH change.[124,125] They all demonstrate that setting involves neutralisation of the polymeric acid by the glass, with associated formation of calcium and aluminium salts. Both infrared spectroscopy and pH determination have shown that neutralisation does not go to completion, but that there remain a few unreacted carboxylic acid groups within the set cement.

Infrared spectroscopy has been useful for the study of the setting reactions, because the expected products of reaction, calcium and aluminium carboxylate units have different carbonyl stretch absorptions, and hence can be detected separately. This has enabled workers to demonstrate that the initial product of reaction is predominantly the calcium salt, and that aluminium polycarboxylate units do not form until some time has elapsed. This is to be expected in that attack of the glass has been shown to favour calcium-rich sites.[108] For some years, this result was interpreted as suggesting preferential release of calcium from the glass. However, Cook reported that aluminium was present at all times throughout the cure process,[126] a finding he arrived at following studies in which he degraded setting cements in 3% KOH solution, and determined the free ions using atomic absorption spectroscopy. This led to the suggestion that aluminium was released from the glass as a condensed species,[127] possibly anionic in nature, and arising from wholesale dissolution of a fraction of the glass. Although not anionic, species of the appropriate degree of condensation, *e.g.* $[Al_{13}O_4(OH)_{24}(H_2O)_{12}]^{7+}$, have been detected in aqueous solution.[128] The release of individual aluminium atoms from such a species might be expected to be slow, and hence the formation of aluminium carboxylate units delayed.

In addition to calcium and aluminium, sodium and silicon have been found to be released from a glass during acid attack,[127] and to be present in the matrix of set cements.[127] This led to the suggestion that these elements were involved in forming an inorganic network of their own, possibly interpenetrating with the metal polyacrylate one. This hypothesis was tested by fabricating cements from a glass and acetic acid. Cements of acetic acid are not immediately insoluble in water, but became so after 6 hours, despite the known solubility of the possible metal acetates.[129] Ageing over a period of six months showed significant increases in compressive strength,[130] a further parallel with the behaviour of properly formulated glass-ionomer cements. Additional support for the hypothesis that this kind of inorganic network may be important came from work of Wilson,[131] who carried out simple experiments in which small amounts of glass were dispersed in either water or aqueous acetic acid, and allowed to stand. In both cases, friable solids were formed on the surface of the liquid, whose composition was related to that of the parent glass. Further evidence to support the existence of this inorganic network has been found in the combined FTIR and MAS-NMR studies of Matsuya *et al.*[132]

However, although there is strong evidence for some sort of wholly inorganic reaction in the setting process of glass-ionomers, its overall importance is not clear. Studies on the effect of molecular weight of the polymer, including on subtle properties such as the relationship between fracture toughness and molecular weight,[133] have shown how important the polymer is in determining the mechanical behaviour of these cements. So far, there has been no attempt to develop an improved quantitative model to explain the observed relationships that also takes account of the possible inorganic network, and hence the case for the involvement of such a network has not been demonstrated to any convincing extent.

Ion-release from Glass-ionomer Cements

A number of ions have been found to be released by glass-ionomers during storage in water or other aqueous media. The main ion of interest is fluoride, as will be considered shortly, but other species have been detected, including aluminium, silicon, sodium and phosphate,[134-136] Surprisingly, little or no calcium is released in neutral conditions,[137] which may suggest that the calcium removed from the glass during setting becomes rapidly bound to polyacrylate molecules, calcium polyacrylate being known to be insoluble in water.

The release of aluminium from glass-ionomers is potentially hazardous,[138] though no adverse effects have been detected for dental grade materials. On the other hand, in experimental materials for bone-contact applications, release of aluminium has been shown to be responsible for deficiencies in the mineralisation of the bone adjacent to the cement.[139] In another, more serious case, an experimental glass-ionomer was used to repair craniofacial defects, and cement came into contact with the cerebro-spinal fluid of patients, who suffered an array of adverse reactions, including stupour, mutism and epilipetic seizures.[140] Since this time (1994), work on these bone augmentation applications seems to have been discontinued,[141] though there has been a notable successful non-dental application, the use of glass-ionomer cement to fabricate preformed ear ossicles.

The release of fluoride from glass-ionomers has been the most studied of all the ion-release processes, and has been shown to be sustained over long periods of time,[142-146] Most of these studies have used water as the medium into which release has been determined, and the few studies there have been have shown that release into human saliva occurs to a much reduced extent.[147,148] Studies have generally indicated that this is beneficial.[149] For example, in one study carried out in the Nordic countries and Australia in the period 1991-1992, 954 questionnaires were completed by dentists, and covered the question of adjacent caries and gingival inflammation in the proximity of dental restorative materials.[150,151] They found that caries around glass-ionomers was reported to occur 'seldom or never' by about 98% of respondents, and 'often' by only about 2%. Conversely, for composite resins, they found that adjacent caries was reported 'seldom or never' by only 26% of respondents, and 'often' by about 70%. It was similar for gingival inflammation. The authors therefore concluded that the use of glass-ionomer cements was beneficial clinically, and that the underlying reason for this positive finding was their fluoride release. Though these conclusions have been challenged,[152] a considerable volume of literature exists to suggests that fluoride release by glass-ionomers is, indeed, beneficial.

Studies of the kinetics of the fluoride release process[153] have shown that it is most closely described by the equation:

$$[F]_c = \frac{[F]_I t}{t_{\frac{1}{2}} + t} + \beta t^{\frac{1}{2}}$$

where $[F]_c$ is the cumulative fluoride release, $[F]_I$ is the total available fluoride for release, t is the equilibration time, $t_{\frac{1}{2}}$ is the time needed to release half of this total amount, and β is a constant.

A comparison of the experimental data and results predicted from this

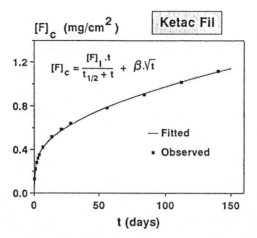

Figure 5.5 *Comparison of calculated and experimental fluoride release from a glass-ionomer cement*

equation is shown in Figure 5.5 for the commercial glass-ionomer cement Ketac Fil (ex. ESPE, Seefeld, Germany). The correlation coefficient for these two lines is 0.9987, showing that the combined hyperbolic and square root dependency is an excellent model of the actual release rate determined by experiment.[154]

This equation shows that fluoride release is the result of two processes. The first, described by the first term on the right hand side of the equation, is an initial short-term elution, which occurs rapidly but which ceases after some time. The second, described by the $\beta t^{\frac{1}{2}}$ term, is a more prolonged process which, as indicated by the square root dependency on time, is a diffusion-based loss of fluoride. Overall, these two terms suggest that fluoride release from these cements is a dissolution–diffusion process, with the relative contributions of these two mechanisms changing over time.

It has been found experimentally that glass-ionomer cements not only release fluoride, but can also absorb fluoride when exposed to an aqueous solution of an appropriate metal fluoride.[155,156] They are thus able to act as reservoirs for fluoride and this capacity means that they can sustain their release of fluoride for many years at clinically useful levels.

The release of fluoride is complicated by the presence of aluminium, with which fluoride forms strong complexes.[157] It is because of this that most studies of fluoride release have employed the decomplexing reagent TISAB to ensure all the fluoride is free and detectable. Nakajima *et al.* determined the amount of both aluminium and fluoride released by a series of glass poly-alkenoates stored in water for up to seven days. There were wide variations in the amount of each of these species, according to which particular brand of glass polyalkenoate was being studied. Fluoride release was generally found to be lower when determined with decomplexing agents, confirming that complexation occurs between aluminium and fluoride. However, by seven days, most of the aluminium had been leached out of the cements, so that after this time, there was no need to employ the decomplexing agent.

One of the modified versions of glass-ionomer cements includes finely-divided metal particles, either in admixture with the glass, or fused to the glass to form a so-called *cermet*.[158] Metal reinforced glass-ionomer cements have been found to release not only fluoride, but also silver and copper in very low amounts.[159] These release studies were carried out using both deionised water and artificial saliva, and showed that, though there was a reasonable level of fluoride release from the cermet-containing cements, it was less than from those cements containing silver alloy only. Very significantly, they found that there were differences in the behaviour of materials in the different aqueous media and, in artificial saliva, release of all ions was reduced.

Glass-ionomer cements not only release ions, they are capable of taking them up again from aqueous solution. In addition to fluoride, whose uptake has already been discussed, glass-ionomers have been shown to take up potassium, sodium and rubidium ions from solutions in which they were stored.[160] The clinical significance of this has not been established, though it suggests a high degree of interaction with aqueous ions and indicates that glass-ionomers are likely to interact with saliva in the mouth, and develop properties that are different from those that have been determined in pure water over many years of experimental study.

Biocompatibility

There have been numerous studies of biocompatibility of glass-ionomer cements, using a variety of cell culture techniques. Three major features of these materials have been identified[161] as leading to their generally good biocompatibility, namely (i) low setting exotherm,[162] (ii) rapid change in pH towards neutral as the cement sets[163] and (iii) release of generally benign ions from the cement. These cements have been found to cause mild irritation of the pulp when used as liners, a feature that probably results from the lower powder:liquid ratio of lining grade cements, which in turn provides less base for reaction with the polyacid.[164] However, this response is no more severe than that from zinc polycarboxylate or zinc phosphate cements.[165]

Resin-modified glass-ionomer cements, which are discussed in detail later, have also been studied for their biocompatibility, though the range of studies has been more limited.[166,167] In cell culture studies against 3T3 mouse fibroblast cells, they have been found to show only limited cytotoxicity, and to be similar to the purely acid–base glass-ionomers.

Clinical Performance

For clinical use, glass-ionomers are provided in a range of formulations, designed for specific uses.[168] These include luting, liners and bases and direct restorations. Materials can be fast or slow setting, to suit the technique of individual dentists, and are provided as powder and liquid combinations for hand mixing, and as capsulated combinations, in two-chamber capsules for automatic mixing. These chambers are separated by a thin membrane, which is broken immediately before use, and the cement mixed using a oscillatory mixing device. These capsules are also provided with a nozzle through which the freshly mixed cement may be extruded at the site of the restoration.

In terms of mechanical properties, hand-mixed and capsule-mixed cements are usually considered equivalent. However, they have been shown to have different degrees of porosity.[169] Hand-mixed and capsule-mixed cements were found to have similar size and distribution of the small (<0.01 mm^3) bubbles introduced by the mixing processes, but hand-mixed cements were found to have a significantly greater number of large diameter bubbles. This may be significant in clinical use and explain the observation that dental prostheses, such as crowns or posts, cemented with capsulated cements have a higher probability of survival than those cemented with hand-mixed materials.

Resin-modified Glass-ionomers

Resin-modified glass-ionomers are restorative materials that contain mainly the components of conventional glass-ionomer cements, *i.e.* basic glass and water-soluble acid, together with organic monomers and their associated initiator systems.[170] The first of these materials was described in a now-abandoned patent application, and consisted of a mixture of composite resin monomers, including bisGMA, plus 2–hydroxyethyl methacrylate, HEMA, as cosolvent.[171] Shortly afterwards, the use of graft copolymers of poly(acrylic acid) was reported.[172] These formulations not only included graft copolymers, but also contained HEMA to act as co-solvent, because the water-solubility of the polymeric acid was compromised by the presence of the wholly organic side-chains.

Whatever the details of their composition, resin-modified glass-ionomers are able to bond to enamel and dentine[173] and also to release fluoride,[174,175] So far, the amount of fluoride release required to produce a therapeutic effect has not been established,[176] but in *in vitro* experiments, resin-modified glass-ionomers were found to have an equivalent caries inhibition to conventional glass-ionomers.[177]

One conseqence of the presence of polyHEMA in the set cements is that they swell in water.[178] This has been considered by a number of authors,[179,180] and shown to vary with the HEMA content of the resin-modified glass-ionomer and also to be influenced, though not simply, by the osmolarity of the storage solution. Water uptake has been shown to follow Fick's law,[181] and diffusion coefficients were found to vary with cure time. Diffusion was found to be more rapid, but to have a higher activation energy, when the external medium was aqueous sodium chloride than when it was pure water, which led to the suggestion that there was some uptake of NaCl, with corresponding conformational change in the polymer, leading to more compact geometry and greater freedom for water molecules to diffuse through the structure. Another study employed confocal microscopy to examine the behaviour of resin-modified

Table 5.7 *Effect of HEMA on a glass-ionomer cement*

Property	Glass-ionomer only	Glass-ionomer + HEMA
Setting time/min	9.1	19.1
Compressive strength/MPa at 24 h	230	147

glass-ionomer placed as wedge-shaped restorations in teeth.[182] This demonstrated that these materials show considerable sensitivity to dehydration, forming distinct gaps at the interface with the tooth within 15 minutes of the onset of dehydration. This led to the conclusion that the addition of HEMA has not significantly improved the susceptibility of the glass-ionomer cement to loss of water from the structure under conditions of low humidity.

The presence of HEMA has been shown to influence the acid–base setting reaction in these materials, making the reaction more sluggish and leading to the formation of a significantly weaker material[183] (see Table 5.7).

HEMA alters the reaction in a variety of ways. It affects the energetics of ionic reactions, since ions are capable of forming less readily in the organic medium, since it has a much lower dielectric constant than water.[184] In addition, organic molecules are known to influence the conformation of the poly(acrylic acid) molecule.[185] Methanol, for example, has similar effects on the strength and setting rate and is known to reduce the radius of gyration of poly(acrylic acid) in solution. Pure water is not a particularly good solvent for poly(acrylic acid), as shown by the Flory–Huggins theta temperature of 14 °C,[186] but it becomes considerably worse as methanol is gradually added to it.

As setting proceeds, there is a tendency for the components of the resin-modified glass-ionomer to undergo phase separation. This is partly driven by the fact that polyHEMA is insoluble in water, a tendency that is reinforced by the increasingly ionic character of the aqueous phase within the cement. Thus, it has been suggested that the ions may show some tendency to cluster into domains that are separate from those containing the organic molecules. However, this has not been demonstrated in set resin-modified glass-ionomer cements, and its consequences in terms of clinical durability are not clear.

Resin-modified glass-ionomers, though, have been widely used within clinical dentistry, and bespoke products are now made for a number of specific tasks. These include restorations, liners and bases, core build-up and luting. Clinical reports for particular applications, such as so-called Class V restorations, *i.e.* those along the gum-line of the anterior teeth, show these materials to be reliable and to give good results.[187] Similarly, their use in paediatric dentistry has been very successful,[188,189] and has led to the suggestion that they should now be used in preference to amalgams in all repairs of the primary dentition.[190]

A number of papers have been concerned with water balance in resin-modified glass polyalkenoates. In one of these studies, Yap and Lee demonstrated that these materials took up significantly more water than composite

Table 5.8 *Shear bond strengths of glass-ionomers
to dentine and enamel*

Substrate	Bond strength/MPa
Dentine	1.5–4.5
Enamel	2.6–9.9

resins, and that this uptake was related to the HEMA content of the materials.[191] They found, though, that there was no correlation between water sorption and solubility, and also that some of the materials bound a proportion of the sorbed water irreversibly. In another study, Nicholson examined the factors that affect water uptake in these materials, using the classical approach of determining diffusion coefficients using Fick's law.[192] He showed that diffusion and solubility depended on the length of the cure time, and that diffusion was more rapid, but had a higher activation energy when the storage medium was a solution of sodium chloride. From this he concluded that the polyelectrolyte molecules undergo a conformational change as aqueous salt solution is absorbed, resulting in more contracted geometry with correspondingly greater freedom for water molecules to diffuse through the structure.[14] Finally, on a related topic, Sidhu *et al.*[193] studied dehydration shrinkage and its relationship to maturity in resin-modified glass polyalkenoates. Using confocal microscopy to examine the behaviour of wedge-shaped cervical cavities, they showed that these materials exhibited considerable sensitivity to dehydration, forming distinct gaps within 15 minutes of the onset of dehydration. Overall, they concluded that the addition of resin has not significantly improved the material's susceptibility to loss of water under conditions of low humidity.

Adhesion to Dentine and Enamel by Glass-ionomer Cements

Glass-ionomer cements have the useful property of inherent adhesion to both dentine and enamel. Bond strengths to these substrates have been determined under a variety of conditions using a range of test methods, and this has resulted in a range of values and failure modes, usually distinguished as adhesive or cohesive, being reported (see Table 5.8).

In general, bond strengths of glass-ionomer cements are greater to enamel than to dentine, leading to the conclusion that bonding occurs to the mineral phase of the tooth, *via* chelation to calcium ions at the surface of the hydroxyapatite.[194] X-Ray photoelectron spectroscopy (XPS) has been used to analyse the interaction of glass-ionomer cement with both enamel and synthetic hydroxyapatite.[195] In both cases the peak representing the carboxyl group of the polyacid was shown to have shifted to a lower binding energy, consistent with chemical reaction to form carboxylate. An estimated 67.5% of total carboxylic acid groups had undergone this reaction, which clearly accounts for the occurrence of true adhesive bonding between glass-ionomers and these substrates. In some ways these findings were surprising, as the surfaces of these substrates are likely to consist of tightly bound water molecules, and any calcium would be expected to be relatively inaccessible. By contrast, for resin-modified glass-ionomers, XPS has not been able to detect any significant ionic bonding at a tooth surface.[196] In a completely different study, adsorption of polyfunctional carboxylic acids to hydroxyapatite was found to be independent of the stability constants of their chelate compounds with calcium, leading to the conclusion that calcium chelate formation alone could not be responsible for adsorption.[197] This suggests that part of the cement's ability to adhere is *via* the formation of hydrogen bonds between the poly(acrylic acid) or polyacrylate ion and individual water molecules, as well as through formation of calcium polyacrylate structures.

Despite the differences observed with XPS, the inclusion of organic monomers appears to make little difference to the measured bond strengths of these materials to the tooth, bond strengths for resin-modified glass-ionomers being at least as great as those for self-hardening types. Moreover, such bonds develop rapidly, so that after 24 hours the strength is of the order of three times greater than it was immediately following placement.

The bonding of glass-ionomers has also been studied using an electron microscopic technique which employs frozen specimens, and therefore immobilises water within the structure.[198] Because of this, there is no desiccation, and hence no cracking or crazing to obscure the images. Consequently, it has been possible to observe quite clearly the zone of interaction that forms between the cement an enamel and dentine (see Figure 5.6). The formation of

Figure 5.6 *Cryostage electron microscope image of a glass-ionomer cement (containing large unreacted glass fragments) in a vital tooth. Note the zone of interaction at the interface*[198]

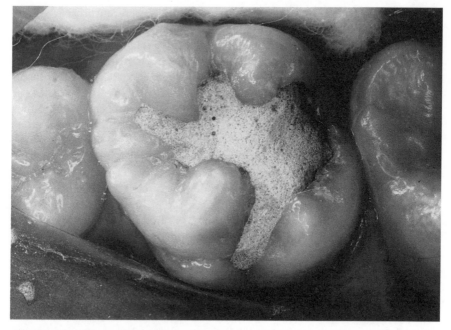

Figure 5.7 *Cermet-containing glass-ionomer nine years after placement*[204]

this zone may explain the observation that long-term retention rates by glass-ionomers are good despite relatively weak shear bonds to non-vital teeth at relatively short time intervals after placement.[199]

Metal-containing Glass-ionomers

An important development in the clinical deployment of glass-ionomer cements has been to include metallic fillers, either bonded to the glass (cermets) or as distinct additional fillers. There are other forms of modified glass-ionomer cement available to the clinician, namely metal-reinforced types. In these materials, conventional glass-ionomers have metal powders incorporated into them, these powders typically being made of the silver–tin alloy used in dental amalgams.[200] Alternatively, this metal can be fused to the glass to form a cermet.[201] However, this process does not seem to enhance the physical properties[202] and reduces the amount of fluoride released to the level where it may be insufficient for clinical benefit.[203] These materials appear to have good durability, at least in patients who are not prone to develop caries, as illustrated in Figure 5.7, where a restoration that has survived for nine years is shown.[204]

Metal reinforced glass-ionomers have also been recommended for use as core build-up materials.[205] They appear to have attractive properties, since they have low coefficients of thermal expansion and good adhesion to the tooth. However, they are more brittle than either amalgam or composite resin, and this tends to limit their success. The evidence is that the best results are obtained when at least two walls of the tooth remain,[206] under which circumstances, the metal-containing glass-ionomers have been found to be as good as amalgam.

5 Polyacid-modified Composite Resins

Polyacid-modified composite resins[207] were first made available to the dental profession in the early 1990s. They are known informally as 'compomers', the term being a contraction of the words *composite* and *ionomer* and designed to indicate their hybrid character. In their original form they were light-activated, cured by free-radical reaction with camphorquinone as initiator, and subsequently made to set using a conventional dental curing lamp that employs blue light (470 nm wavelength) in the same way as true composite resins.

Polyacid-modified composite resins differ from conventional composites in that one of the components is a monomer bearing a few carboxylic acid groups. There are insufficient of these to make the molecule water soluble, so that these materials contain no water and do not set by neutralisation. Instead they set entirely by addition polymerisation. Then, once fully set, they are able to draw in water from the saliva to initiate a small amount of neutralisation of the monomer by the basic glass filler

A major advantage of these materials is their fluoride release.[208] However, such release is significantly less than for glass-ionomers, though there has been no evidence published to date that suggests the development of secondary caries is a problem with these materials.

Other advantages of compomers are that they have excellent aesthetics, and are extremely easy to use. Their viscosity and general handling characteristics seem better than those of conventional composites, though it is too soon for their durability to have been assessed fully. Initial results are promising, for example in so-called Class V cavities[209] and for time periods of the order of a year,[210] but longer term data are not yet available. Unlike glass-ionomers, they do not show ion-exchange bonding to teeth, but instead have to be bonded in place with techniques similar to those of conventional composites. This means acid-etching to enamel and the use of an acetone-based primer for dentine.

For prosthodontic applications, there are also compomer luting cements. These are two-paste materials which are based on the older chemical-cure initiator systems used in composite resins, a feature which allows them to become fully cured beneath opaque devices such as metal crowns.

The original materials were found to have relatively poor wear characteristics. For example, one study showed them to lose 100 μm from their surface in just six months,[211] a figure which was roughly three times that of a composite resin designed for similar clinical use.[212] This has not proved to be a fundamental weakness of these materials, however, and recently formulations have become available which show wear rates that are similar to those of current composite resins.[213]

6 Aesthetic Dental Restoratives: General Considerations

The use of adhesive materials in dentistry is growing rapidly at the present time.[214] Adhesive materials and techniques are being employed because not only are they aesthetic, matching the tooth in colour and translucency, but also they are conservative: less of the structure has to be cut away in preparing the cavity for these materials than has conventionally been necessary for amalgams. This change towards the use of adhesive materials has also affected the way in which amalgams are being used, since there are now special bonding agents designed for use with amalgam restorations. These bonding agents are designed to eliminate leakage at the margins of amalgam restorations, though at the price of an increase in the difficulty of the placement technique.

In general, aesthetic materials are found to be less durable in clinical service than amalgams and several factors have been identified as influencing the longevity of such restoratives.[215] These include:

(i)   Caries, *i.e.* secondary caries which may develop around the restoration, destroying more of the dentine or enamel than was previously affected and which requires replacement of the original restoration;
(ii) Mechanical failure of the restoration. This may include the opening of margins between the restoration and the tooth or fracture of the filling.
(iii) Mechanical failure of the tooth. This arises in certain circumstances because the placement of a filling weakens a tooth, and where this has happened to a critical extent the normal forces experienced during mastication can cause the cusps of the tooth to break off.
(iv) Further treatment of the tooth is required for clinical reasons; for example there may be a need to fill the root canal or to remove the tooth because of gum disease.

This list does not specifically include adhesive failure of the restoration, though this could be included under the heading of 'Mechanical failure of the restoration'. It also does not specifically include wear, though this is a major area of concern for non-amalgam fillings, all of which show inferior wear properties compared with either amalgam or the tooth itself. Materials failure has been suggested as being responsible for an estimated 23% of restoration failures,[216] the rest being attributed to clinical factors, such as secondary caries, or to poor placement techniques. Thus, although restorations are known to have limited lifetimes, and many studies on durability have been reported over the years, failure in service rarely, if ever, seems to be primarily due to loss of adhesion. Other factors always seem to predominate.

A variety of failure mechanisms have been identified for aesthetic materials in clinical service. For example, glass-ionomers are brittle and liable to fracture in load-bearing applications. Composite resins and glass-ionomers of all types may discolour, either through surface staining as they become rougher in service, or through water-uptake, leading to changes in translucency.[217-220]

Table 5.9 *Lifetime estimates of restorative dental materials*

Materials	Survival quartiles (years)		
	75th	50th	25th
Amalgams	10.52	22.52	43.00
Composite resins	7.91	16.72	34.41
Glass-ionomers	11.25	–	–

For glass-ionomers, there were too few failures during the study period to obtain the median survival time.

Wear is also important as a failure mechanism for composite resins, particularly those used to restore molar teeth.[221]

Glass-ionomers are significantly less durable *in vivo* than amalgams, as shown in numerous studies, even where the failure mechanisms have not been identified. For example, in primary teeth, a study of longevity and cariostatic effects found overall that there was considerable loss of glass-ionomer, whereas amalgam was more or less unchanged.[222] Glass-ionomers showed reduced caries progression compared with the amalgams, but this did not compensate for the loss of material.

The reasons for the replacement of direct restorations by dental practitioners have been reported in recent studies,[223,224] the former being based on responses by dentists in the USA, the latter by dentists in the UK. For glass-ionomers the main reason for replacement was given as secondary caries. Secondary caries to this extent had never previously been reported for these materials, and one explanation suggested is that glass-ionomers may be placed in more challenging situations than other materials, *i.e.* in patients more prone to caries or with lower standards of oral hygiene. The alternative possibility is that a number of materials have been marketed over recent years as 'glass-ionomers' but that they are not necessarily proper glass-ionomers, with the properties of acid-base setting, ion-exchange adhesion and fluoride release.[225] It may be that this confusion due to marketing tactics led to some of the materials in the original study being wrongly identified. Whatever the reason for the findings, they were at odds with many other studies of these materials. For example, another recent clinical study,[226] this time based on experience in Australia, reported on the survival of restorations and reported results for glass-ionomers that were particularly good (see Table 5.9). In particular, there had been too few failures to allow calculation of their median survival time. These results are among the best that have been reported, possibly due to the fact that the patients were well-motivated and attended the dentist regularly. This illustrates one further difficulty in this field, namely that the quality of care that patients give can alter the longevity of the restorations.

7 Prosthetic Materials

The main material used in prosthetics, *i.e.* to replace the teeth in the form of full or partial dentures, is poly(methyl methacrylate). This substance is polymerised to form the prosthesis by what is known as the *dough* technique. In this technique, polymerised methyl methacrylate is added to unpolymerised liquid monomer in the ratio 3:1 by volume (approximately 2.5:1 by mass), and the monomer polymerised, usually as a result of interaction with free radical initiator contained in the polymer powder.[227] Because only about one-third of the mixture is monomer, the overall contraction on polymerisation is much reduced, and the resulting prosthesis is more readily fabricated to the necessary dimensions for the individual patient.

In general, the properties of poly(methyl methacrylate) denture bases are satisfactory for service. They have adequate strength, though they are known to flex under conditions of use, and may fracture down the mid-line.[228] Generally, these materials contain small amounts of free monomer and, over time, these can leach into the mouth and cause irritation. Irritation can also be caused in patients who have a natural intolerance to hard denture bases, and in patients whose soft tissues need to heal following oral surgery, tooth extraction or inflammation.

The latter patients can have temporary soft materials placed as liners for their dentures. These are typically longer chain acrylic polymers, such as poly(ethyl methacrylate), to which ethanol and an ester, such as butyl phthalyl glycol glycolate, are added as plasticisers.[229] These are able to retain their resilience for relatively short times only, because they tend to leach out the plasticiser and become more rigid.[230] However, this is sufficient for the patients who need them to have experienced healing of the gingival tissue.

More permanent soft linings have been fabricated from a variety of inherently resilient polymers. Of these, the most successful has been a heat-polymerised silicone,[231] though other silicones, for example of the condensation-type, have also been studied experimentally.[232] These materials tend to retain their mechanical properties for much longer than the tissue conditioners, but even they have a tendency to harden and lose resilience with time. Research is continuing into the problem of fabricating durable soft lining materials.

8 Endodontic Materials

Endondontics is that branch of dentistry concerned with filling root canals of teeth from which it has been necessary to remove the pulps. This means that the repaired tooth is dead, but its retention for function and space-filling is clinically desirable. From the surgical point of view, the critical part of endodontics is to create an impermeable seal at the apex of the tooth, and thus prevent the tooth becoming a seat of infection.

In order to create such a seal, following surgical excision of the pulp and the widening of the canals of the tooth with fine graded files, preformed cones of gutta percha are inserted, and retained in place with some kind of cement, typically zinc oxide–eugenol, but also possibly an epoxy.

Gutta percha has been the material of choice for this application for several years.[233] It is reasonable pliable, and can be deformed to fit the curvature of natural root canals. If necessary, its properties in this respect can be improved by adding small amounts of organic solvent, to act effectively as plasticiser and assist deformation into the shape of the root canal.[234] In the past silver points were used as an alternative, but this has almost entirely ceased, due in part to the very poor biocompatibility of silver in a region close to the periodontal ligament.

The gutta percha points, even when solvent-softened, cannot form an impermeable seal with the root canal, as they lack sufficient flow and flexibility. The seal is completed by placing them with a cement or sealer. Most widely used is zinc oxide–eugenol,[235] whose chemistry was described in an earlier section of this chapter. It has an advantage in that eugenol has slight bactericidal properties, and hence the use of this material can ensure that the newly cleaned root canal remains sterile.

Alternatively, epoxy resins can be used.[236] There may be biocompatibility problems with these materials should they be expressed beyond the apex of the tooth, though clinicians tend to try to avoid this. Also, endodontic grade materials are formulated with high molecular weight amine curing agents, and these tend to lack mobility, and hence hardly leach at all from the set or setting material, thereby minimising damage to the surrounding tissue.

9 Orthodontic Materials

Orthodontics is that branch of dentistry concerned with mechanical realignment of the teeth.[237] As such, it uses a variety of materials which are fabricated into special devices. The main ones are (i) brackets, usually retained either mechanically or by adhesion to the surface of the individual teeth, and (ii) wires, designed to exert force to the teeth, and draw them into the correct orientation. Brackets are typically made of stainless steel, usually type 316L austenitic stainless steel. This generally performs well, though may occasionally show signs of corrosion in the mouth.[238] More aesthetically pleasing brackets have been fabricated from polycarbonate,[239] from high-purity alumina[240] and from yttria-stabilised zirconia.[241] They are bonded by techniques and materials previously described for restorative materials, *i.e.* acid etch plus use of bisGMA/TEGDMA blends,[242] or using cements such as zinc polycarboxylate or glass-ionomer.[243]

Four types of wire are available for use in orthodontics, namely austenitic stainless steel, Co-Cr-Ni (Elgiloy Blue), β-titanium and Ni-Ti, so-called *nitinol*. Properties vary, and there is no one type that is preferred for clinical use.[244] Stainless steel and Elgiloy Blue alloys are relatively inexpensive, making wires fabricated from these materials relatively cheap. However, the advantage of the superior tensile properties of nitinol wires is somewhat undermined by the presence of nickel within them, and their slight but distinct ability to cause a nickel sensitisation reaction in certain patients.

10 References

1   S. Hojo, M. Komatsu, R. Okuda and N. Takahashi, *J. Dent. Res.*, 1994, **73**, 1853.
2   S. Hojo, N. Takahashi and T. Yamada, *J. Dent. Res.*, 1991, **70**, 182.

3   R. M. Stephan, *J. Am. Dent. Assoc.*, 1940, **27**, 718.
4   A. Banerjee, E. A. M. Kidd and T. F. Watson, *J. Dent.*, 2000, **28**, 179.
5   M. Brannström, G. Johnson and J. Friskopp, *J. Dent. Children*, 1980, **47**, 46.
6   N. Nakabayashi, M. Ashizawa and M. Nakamura, *Quintessence Int.*, 1992, **23**, 135.
7   A. D. Wilson and J. W.McLean, *Glass Ionomer Cement*, Quintessence Books, Chicago, IL, 1988.
8   J. W. McLean, J. W. Nicholson and A. D. Wilson, *Quintessence Int.*, 1994, **25**, 587.
9   J. F. McCabe, *Biomaterials*, 1998, **19**, 521.
10 R. L. Bowen, US Patent 3 066 122, 1962.
11 M. Braden and K. W. M. Davy, *Biomaterials*, 1986, **7**, 474.
12 K. Anusavice, *Science of Dental Materials*, 10th Edition, W. B. Saunders, Philadelphia, PA, 1996.
13 D. C. Watts, O. Amer and E. Combe, *Br. Dent. J.*, 1984, **156**, 209.
14 P. Dionysopoulos and D. C. Watts, *J. Oral. Rehabil.*, 1990, **17**, 9.
15 H. F. Gruber, *Prog. Polym. Sci.*, 1992, **17**, 953.
16 E. Andrzejewska, L.-A. Linden and J. F. Rabek, *Macromol. Chem. Phys.*, 1998, **199**, 441.
17 B. M. Monroe and S. A. Weiner, *J. Am. Chem. Soc.*, 1969, **91**, 450.
18 M. B. Rubin and Z. Hershtik, *J. Chem. Soc., Chem. Commun.*, 1970, 1267.
19 J. Meinwald and H. O. Klingele, *J. Am. Chem. Soc.*, 1966, **88**, 2071.
20 C. Decker and K. Moussa, *J. Appl. Polym. Sci.*, 1987, **34**, 1603.
21 K. Inomata, Y. Minoshima, T. Matsumoto and K. Tokumaru, *Polym. J.*, 1993, **25**, 1199.
22 M. Schwarzenberg, J. Sperling and D. Elad, *J. Am. Chem. Soc.*, 1973, **95**, 6418.
23 D. Sustercic, P. Cevc, N. Funuk and M. M. Pintar, *J. Mater. Sci., Mater. Med.*, 1997, **8**,
24 G. F. Cowperthwaite, J. J. Foy and M. A. Malloy, in *Polymer Science and Technology*, Vol. 14, ed. C. G. Gebelein and F. F. Koblitz, Plenum Press, New York, 1981.
25 J. L. Ferracane, PhD thesis, Northwestern University, Chicago, IL, 1983. Quoted in R. L. Clarke, Chapter 2 in *Polymeric Dental Materials*, ed. M. Braden, R. Clarke, J. Nicholson and S. Parker, Springer-Verlag, Heidelberg, 1997.
26 F. Lutz and R. W. Phillips, *J. Prosthet. Dent.*, 1983, **50**, 480.
27 H. R. Stanley, R. L. Bowen and J. Folio, *J. Dent. Res.*, 1979, **58**, 1507.
28 K. D. Jorgensen, P. Horsted, O. Janum, J. Krogh and J. Schulz, *Scand. J. Dent. Res.*, 1979, **87**, 140.
29 F. Lutz and R. W. Phillips, *J. Prosthet. Dent.*, 1983, **50**, 480.
30 T. K. Kwei, *J. Polym. Sci.*, 1965, **3**, 3229.
31 E. P. Plueddemann, *Silane Coupling Agents*, Plenum, New York, 1982.
32 K. J. Soderholm, *J. Dent. Res.*, 1981, **60**, 1867.
33 M. G. Buonocore, *J. Dent. Res.*, 1955, **34**, 849.
34 L. R. Legler, D. H. Retief, E. L. Bradley, F. R. Denys and P. L. Sadowsky, *Am. J. Orthodont. Dentofacial Othop.*, 1989, **96**, 485.
35 D. H. Pashley and R. M. Carvalho, *J. Dent.*, 1997, **25**, 355.
36 W. H. Douglas, *J. Dent.*, 1989, **17**, 209.
37 S. A. M. Ali and D. F. Williams, *Clin. Mater.*, 1993, **14**, 243.
38 D. H. Retief, *Int. Dent. J.*, 1994, **44**, 19.
39 E. J. Swift, J. Perdigao and H. O. Heymann, *Quintessence Int.*, 1995, **26**, 95.
40 S. A. M. Ali, J. Jaworzyn and D. F. Williams, *J. Adhesion Sci. Technol.*, 1990, **4**, 79.

41 N. Nakabayashi, K. Kojima and E. T. Masuhara, *J. Biomed. Mater. Res.*, 1982, **16**, 265.

42 D. H. Pashley. *Trans. Acad. Dent. Mater.*, 1990, **3**, 55.

43 T. Wang and N. Nakabayashi, *J. Dent. Res.*, 1991, **70**, 59.

44 B. van Meerbek, A. Dhem, M. Goret-Nicaise, M. Braem, P. Lambrechts and G. Vanherle, *J. Dent. Res.*, 1993, **72**, 495.

45 C. L. Davidson, A. I. Abdalla and A. J. de Gee, *J. Oral Rehabil.*, 1993, **20**, 291.

46 P. R. Walshaw and D. M. McComb, *J. Dent.*, 1995, **23**, 281.

47 M. Nakabayashi, M. Nakamura and N. Yasuda, *J. Esthet. Dent.*, 1991, **3**, 133.

48 N. Nakabayashi, E. Kojim and E. Masuhara, *J. Biomed. Mater. Res.*, 1992, **16**, 265.

49 N. Nakabayashi, M. Ashizawa and M. Nakamura, *Quintessence Int.*, 1992, **23**, 135.

50 M. Ferrari, G. Goracci and F. Garcia-Godoy, *Am. J. Dent.*, 1997, **10**, 224.

51 J. W. Nicholson and G. Singh, *Biomaterials*, 1997, **17**, 2023.

52 H. A. Alhadainy and A. I. Abdalla, *Am. J. Dent.*, 1996, **9**, 77.

53 C. Holderegger, J. P. Stefan and H. Luthy, *Am. J. Dent.*, 1997, **10**, 71.

54 R. van Noort, *J. Dent.*, 1998, **26**, 195.

55 U. B. Fritz, W. Finger and H. Stean, *Quintessence Int.*, 1998, **29**, 567.

56 J. M. Powers, W. J. Finger and J. Xie, *J. Prosthodont.*, 1995, **4**, 28.

57 A. R. Grieve, W. P. Saunders and A. H. Alani, *J. Oral Rehabil.*, 1993, **20**, 11.

58 E. A. Kidd, F. Toffenetti and I. A. Mjör, *Int. Dent. J.*, 1992, **42**, 127.

59 T. Tatasku, B. Ciucchi, J. A. Horner, W. G. Matthews and D. H. Pashley, *Oper. Dent.*, 1995, **20**, 15.

60 H. Li, M. F. Burrow and M. J. Tyas, *J. Adhesive Dent.*, 2000, **2**, 57.

61 R. W. Bryant and K.-L. V. Hodge, *Aust. Dent. J.*, 1994, **39**, 77.

62 S. N. White and V. Kipnis, *J. Prosthet. Dent.*, 1993, **69**, 28.

63 A. L. Rochette, *J. Prosthet. Dent.*, 1973, **30**, 418.

64 G. J. Livaditis, *J. Am. Dent. Assoc.*, 1980, **110**, 926.

65 G. J. Livaditis and V. P. Thompson, *J. Prosthet. Dent.*, 1982, **47**, 52.

66 N. N. Greenwood and A. Earnshaw, *The Chemistry of the Elements*, Pergamon, Oxford, UK, 1984.

67 A. D. Wilson, B. E. Kent and B. G. Lewis, *J. Dent. Res.*, 1970, **49**, 1049.

68 H. K. Worner and A. R. Docking, *Aust. Dent. J.*, 1958, **3**, 215.

69 A. D. Wilson, G. Abel and B. G. Lewis, *Br. Dent. J.* 1974, **137**, 313.

70 S. Crisp, I. K. O'Neill, H. J. Prosser, B. Stuart and A. D. Wilson, *J. Dent. Res.* 1978, **57**, 245.

71 J. Margerit, B. Cluzel, J. M. Leloup, J. Nurit, B. Pauvert and A. Terol, *J. Mater. Sci., Mater. Med.* 1996, **7**, 623.

72 A. D. Wilson, G. Abel and B. G. Lewis, *J. Dent.* 1976, **4**, 28.

73 A. D. Wilson and J. W. Nicholson, *Acid-Base Cements*, Cambridge University Press, Cambridge, UK, 1993.

74 J. F. Flagg, *Dent. Cosmos*, 1875, **27**, 465.

75 G. M. Brauer, H. Argenter and G. Durany, *J. Res. Natl. Bur. Stand. (U.S.)*, 1964, **68A**, 619.

76 A. D.Wilson and R. J. Mesley, *J. Dent. Res.*, 1972, **51**, 1581.

77 A. D. Wilson and R. F. Batchelor, *J. Dent. Res.*, 1970, **49**, 593.

78 T. D. Gilson and G. E. Myers, *J. Dent. Res.*, 1970, **49**, 14.

79 A. D. Wilson, *J. Biomed. Mater. Res.*, 1982, **16**, 549.

80 J. W. Nicholson, P. J. Brookman, O. M. Lacy, G. S. Sayers and A. D. Wilson, *J. Biomed. Mater. Res.*, 1988, **22**, 623.

81   J. M. Paddon and A. D. Wilson, *J. Dent.*, 1976, **4**, 183.

82   R. G. Hill and S. Labok, *J. Mater. Sci.*, 1991, **26**, 67.

83   J. W. Nicholson, S. J. Hawkins and E. A. Wasson, *J. Mater. Sci., Mater. Med.*, 1993, **4**, 32.

84   J. W. Nicholson and A. D. Wilson, *J. Mater. Sci., Mater. Med.*, 2000, **11**, 357.

85   A. O. Akinmade and R. G. Hill, *Biomaterials*, 1992, **13**, 931.

86   A. O. Akinmade and J. W. Nicholson, *Biomaterials*, 1995, **16**, 149.

87   A. D. Wilson and J. W. McLean, *Glass Ionomer Cement*, Quintessence Books, Chicago, IL, 1988.

88   W. Schmitt, R. Purrmann, P. Jochum and O. Gasser, Eur. Pat. Appl. 24 056, 1981.

89   J. Ellis and A. D. Wilson, *J. Mater. Sci. Lett.*, 1990, **9**, 1058,

90   J. W. Nicholson, *Biomaterials*, 1998, **19**, 485.

91   G. J. Mount, *Colour Atlas of Glass-ionomer Cements*, 2nd Edition, Dunitz, London, 1994.

92   M. Mandel in *Polyelectrolytes*, ed. M. Hara, Marcel Dekker, New York, 1993.

93   T. Kitano, A. Taguchi, I. Noda and M. Nagasawa, *Macromolecules*, 1980, **13**, 57.

94   E. A. Wasson, PhD thesis, Brunel University, 1992. Quoted in J. W. Nicholson, *Biomaterials*, 1998, **19**, 485.

95   F. T. Wall and J. W. Drennan, *J. Polym. Sci.*, 1951, **7**, 83.

96   F. Oosawa, *Polyelectrolytes*, Marcel Dekker, New York, 1970.

97   H. P. Gregor and M. Frederick, *J. Polym. Sci.*, 1957, **23**, 477.

98   U. P. Strauss and Y. P. Leung, *J. Am. Chem. Soc.*, 1965, **87**, 1476.

99   A. J. Begala and U. P. Strauss, *J. Phys. Chem.*, 1972, **76**, 254.

100 R. Rymden and P. Stilbs, *J. Phys. Chem.*, 1985, **89**, 2425.

101 R. Rymden and P. Stilbs, *J. Phys. Chem.*, 1985, **89**, 3502.

102 J. W. Nicholson, *J. Appl. Polym. Sci.*, 2000, **78**, 1680.

103 A. Ikegami, *J. Polym. Sci.*, 1964, **A2**, 907.

104 A. Ikegami, *Biopolymers*, 1968, **6**, 431.

105 R. G. Hill and A. D. Wilson, *Glass Technol.*, 1988, **29**, 150.

106 A. D. Wilson, S. Crisp, H. J. Prosser and S. A. Merson, *Ind. Eng. Chem. Prod. Res.*, 1980, **19**, 263.

107 A. D. Neve, V. Piddock and E. C. Combe, *Clin. Mater.*, 1992, **9**, 13.

108 M. Darling and R. G. Hill, *Biomaterials*, 1994, **15**, 289.

109 S. Deb and J. W. Nicholson, *J. Mater. Sci., Mater. Med.*, 1999, **10**, 471.

110 W. H. Zacheriasen, *J. Am. Chem. Soc.*, 1932, **54**, 3841.

111 D. J. Wood amd R. G. Hill, *Clin. Mater.*, 1991, **7**, 301.

112 E. De Barra and R. G. Hill, *Biomaterials*, 1998, **19**, 495.

113 A. D. Wilson, S. Crisp and B. G. Lewis, *J. Dent.*, 1977, **5**, 117.

114 S. Crisp, B. G. Lewis and A. D. Wilson, *J. Dent.*, 1977, **5**, 51.

115 S. Crisp, B. G. Lewis and A. D. Wilson, *J. Dent.*, 1976, **4**, 287.

116 A. D. Wilson, S. Crisp and A. J. Ferner, *J. Dent. Res.*, 1976, **55**, 489.

117 J. A. Williams and R. W. Billington, *J. Oral Rehabil.*, 1991, **18**, 163.

118 G. J. Pearson and A. S. Atkinson, *Biomaterials*, 1991, **12**, 658.

119 S. Crisp, M. A. Pringuer, D. Wardleworth and A. D. Wilson, *J. Dent. Res.*, 1974, **53**, 1414.

120 J. W. Nicholson, P. J. Brookman, O. M. Lacy and A. D. Wilson, *J. Dent. Res.*, 1988, **67**, 1451.

121 H. J. Prosser, C. P. Richards and A. D. Wilson, *J. Biomed. Mater. Res.*, 1982, **16**, 431.

122 T. I. Barry, D. J. Clinton and A. D. Wilson, *J. Dent. Res.*, 1979, **58**, 1072.

123 P. Hatton and I. M. Brook, *Br. Dent. J.*, 1992, **173**, 275.
124 D. C. Smith and N. D. Ruse, *J. Am. Dent. Assoc.*, 1986, **112**, 654.
125 E. A. Wasson and J. W. Nicholson, *J. Dent.*, 1993, **21**, 122.
126 W. D. Cook, *J. Biomed. Mater. Res.*, 1983, **17**, 1015.
127 E. A. Wasson and J. W. Nicholson, *Br. Polym. J.*, 1990, **23**, 179.
128 D. N. Waters and M. S. Henty, *J. Chem. Soc., Dalton Trans.*, 1977, 243.
129 E. A. Wasson and J. W. Nicholson, *Clin. Mater.*, 1991, **7**, 289.
130 E. A. Wasson and J. W. Nicholson, *J. Dent. Res.*, 1993, **72**, 481.
131 A. D. Wilson, *J. Mater. Sci. Lett.*, 1996, **15**, 275.
132 S. Matsuya, T. Maeda and M. Ohta, *J. Dent. Res.*, 1996, **75**, 1920
133 R. G. Hill, A. D. Wilson and C. P. Warrens, *J. Mater. Sci.*, 1989, **24**, 363.
134 S. Crisp, B. G. Lewis and A. D. Wilson, *J. Dent. Res.*, 1976, **55**, 1032.
135 A. D. Wilson, D. M. Groffman and A. T. Kuhn, *Biomaterials*, 1985, **6**, 378.
136 M. S. Bapnas and H. J. Mueller, *J. Oral Rehabil.*, 1994, **21**, 577.
137 S. Crisp, B. G. Lewis and A. D. Wilson, *J. Dent.*, 1980, **8**, 68.
138 J. W. Nicholson, J. H. Braybrook and E. A. Wasson, *J. Biomater. Sci., Polym. Ed.*, 1991, **2**, 277.
139 D. H. Carter, P. Sloan, I. M. Brook and P. V. Hatton, *Biomaterials*, 1997, **18**, 459.
140 J. L. Renard, D. Felton and D. Bequet, *Lancet*, 1994, **344**, 8914.
141 J. W. Nicholson, *Proc. Inst. Mech. Eng., Part H*, 1998, **212**, 121.
142 M. Cranfield, A. T. Kuhn and G. B. Winter, *J. Dent.*, 1982, **10**, 333.
143 A. D. Wilson, D. M. Groffman and A. T. Kuhn, *Biomaterials*, 1985, **6**, 431.
144 R. J. G. De Moor, R. M. H. Verbeeck and E. A. P. De Maeyer, *Dent. Mater.*, 1996, **12**, 88.
145 B. F. El Mallakh and N. K. Sarkar, *Dent. Mater.*, 1990, **6**, 118.
146 L. Forsten, *Scand. J. Dent. Res.*, 1990, **98**, 179.
147 F. Rezk-Lega, B. Ogaard and G. Rölla, *Scand. J. Dent. Res.*, 1991, **99**, 60.
148 A. Bell, S. L. Creanor, R. H. Foye and P. Saunders, *J. Oral Rehabil.*, 1999, **26**, 407.
149 M. J. Tyas, *Aust. Dent. J.*, 1991, **36**, 236.
150 L. Forsten, *Acta Odontol. Scand.*, 1993, **51**, 195.
151 L. Forsten, G. M. Knight and G. J. Mount, *Aust. Dent. J.*, 1994, **39**, 339.
152 I. Mjör, *Quintessence Int.*, 1996, **27**, 171.
153 R. J. G. De Moor, R. M. H. Verbeeck and E. A. P. De Maeyer, *Dent. Mater.*, 1996, **12**, 88.
154 R. M. H. Verbeeck, E. A. P. De Maeyer, L. A. M. Marks, R. J. G. De Moor, A. M. J. C. De Witte and L. M. Trimpeneers, *Biomaterials*, 1998, **19**, 509.
155 S. L. Creanor, L. M. C. Carruthers, W. P. Saunders, R. Strang and R. H. Foye, *Caries Res.*, 1994, **28**, 322.
156 A. M. Diaz-Arnold, D. C. Holmes. D. W. Wistrom and E. J. Swift, *Dent. Mater.*, 1995, **11**, 96.
157 H. Nakajima, H. Komatsu and T. Okabe, *J. Dent.*, 1997, **25**, 137.
158 E. A. Wasson, *Clin. Mater.*, 1993, **12**, 181.
159 J. A. Williams, R. W. Billington and G. J. Pearson, *J. Oral Rehab.*, 1997, **24**, 369.
160 P. C. Hadley, R. W. Billington and G. J. Pearson, *Biomaterials*, 1999, **20**, 891.
161 J. W. Nicholson, J. H. Braybrook and E. A. Wasson, *J. Biomater. Sci., Polym. Ed.*, 1991, **2**, 277.
162 S. Crisp, M. A. Jennings and A. D. Wilson, *J. Oral Rehabil.*, 1978, **5**, 139.
163 W. D. Cook, *Biomaterials*, 1982, **3**, 232.
164 D. C. Smith and N. D. Ruse, *J. Am. Dent. Assoc.*, 1986, **112**, 654.

165 R. J. Heys, M. Fitzgerald, D. R. Heys and G. T. Charbeneau, *J. Am. Dent. Assoc.*, 1987, **114**, 607.
166 P. Sasanaluckit, K. R. Albustany, P. J. Doherty and D. F. Williams, *Biomaterials*, 1993, **14**, 906.
167 K. C. Kan, L. B. Messer and H. H. Messer, *J. Dent. Res.*, 1997, **76**, 1502.
168 G. J. Mount, *Biomaterials*, 1998, **19**, 573.
169 C. A. Mitchell and W. H. Douglas, *Biomaterials*, 1997, **18**, 1127.
170 J. W. McLean, J. W. Nicholson and A. D. Wilson, *Quintessence Int.*, 1994, **25**, 587.
171 J. M. Antonucci, J. E. McKinney and R. W. Stansbury, US Patent application No. 160 858, 1988.
172 S. B. Mitra, Eur. Pat. Appl. No. 0 323 120 A2, 1989.
173 S. B. Mitra, *J. Dent. Res.*, 1991, **70**, 72.
174 S. B. Mitra, *J. Dent. Res.*, 1991, **70**, 75.
175 H. Forss, *J. Dent. Res.*, 1993, **72**, 1257.
176 J. F. McCabe, *Biomaterials*, 1998, **19**, 521.
177 H. Ulukapi, Y. Benderli and M. Soyman, *J. Oral Rehabil.*, 1996, **23**, 197.
178 J. W. Nicholson, H. M. Anstice and J. W. McLean, *Br. Dent. J.*, 1992, **173**, 98.
179 I. C. B. Small, T. F. Watson, A. V. Chadwick and S. K. Sidhu, *Biomaterials*, 1998, **19**, 545.
180 A. Yap and C. M. Lee, *J. Oral Rehabil.*, 1997, **24**, 310.
181 J. W. Nicholson, *J. Mater. Sci.; Mater. Med.*, 1997, **8**, 691.
182 S. K. Sidhu, M. Sherriff and T. F. Watson, *J. Dent. Res.*, 1997, **76**, 1495.
183 H. M. Anstice and J. W. Nicholson, *J. Mater. Sci.; Mater. Med.*, 1994, **5**, 299.
184 J. W. Nicholson and H. M. Anstice, *J. Mater. Sci.; Mater. Med.*, 1994, **5**, 119.
185 N. T. M. Klooster, F. van der Trouw and M. Mandel, *Macromolecules*, 1984, **17**, 2087.
186 P. Molyneux, in *Water: a Comprehensive Treatise*, ed. F. Franks, Plenum Press, New York, 1975.
187 M. Tyas, *Aust. Dent. J.*, 1995, **40**, 167.
188 T. P. Croll, *Quintessence Int.*, 1993, **24**, 109.
189 J. W. Nicholson and T. P. Croll, *Quintessence Int.*, 1997, **28**, 705.
190 T. P. Croll and C. M. Killian, *Quintessence Int.*, 1992, **24**, 723.
191 A. Yap and C. M. Lee, *J. Oral Rehab.*, 1997, **24**, 310.
192 J. W. Nicholson, *J. Mater. Sci.; Mater. Med.*, 1997, **8**, 691.
193 S. K. Sidhu, M. Sherriff and T. F. Watson, *J. Dent. Res.*, 1997, **76**, 1495.
194 A. D. Wilson, D. R. Powis and H. J. Prosser, *J. Dent. Res.*, 1983, **62**, 590.
195 Y. Yoshida, B. Van Meerbeek, Y. Nakayama, J. Snauwaert, L. Hellemans, P. Lambrechts, G. Vanherle and K. Waskasa, *J. Dent. Res.*, 2000, **79**, 709.
196 A. Lin, N. S. McIntyre and R. S. Davidson, *J. Dent. Res.*, 1992, **71**, 1836.
197 J. C. Skinner, H. J. Prosser, R. P. Scott and A. D. Wilson, *Biomaterials*, 1986, **7**, 438.
198 H. Ngo, G. J. Mount and M. C. R. B. Peters, *Quintessence Int.*, 1997, **28**, 63.
199 M. J. Tyas and D. R. Beech, *Aust. Dent. J.*, 1985, **30**, 260.
200 J. J. Simmons, *Texas Dent. J.*, 1983, **100**, 6
201 J. W. McLean and O. Gasser, *J. Am. Dent. Assoc.*, 1990, **120**, 43.
202 E. A. Wasson, *Clin. Mater.*, 1993, **12**, 181.
203 L. Forsten, in *Glass-ionomers: The Next Generation*, ed. P. Hunt, International Symposia in Dentistry PC, Philadelphia, PA, 1994.
204 J. W. Nicholson and T. P. Croll, *Quintessence Int.*, 1997, **28**, 705.
205 E. A. Wasson, *Clin. Mater.*, 1993, **12**, 181.

206 J. P. de Wald, C. J. Arcoria and J. L. Ferracane, *Dent. Mater.*, 1990, **6**, 129.
207 J. W. McLean, J. W. Nicholson and A. D. Wilson, *Quintessence Int.*, 1994, **25**, 587.
208 B. J. Millar, F. Abiden and J. W. Nicholson, *J. Dent.*, 1998, **26**, 133.
209 U. Blunck and J. F. Roulet, *J. Adhesive Dent.*, 1999, **1**, 143.
210 R. J. Crisp and F. J. T. Burke, *Quintessence Int.*, 2000, **31**, 181.
211 M. C. R. B. Peters, F. J. M. Roeters and F. W. A. Frankenmolen, *Am. J. Dent.*, 1996, **9**, 83.
212 K. M. Y. Hse and S. H. Y. Wei, *J. Am. Dent. Assoc.*, 1997, **128**, 1088.
213 Y. Luo, E. C. M. Luo, D. T. S. Fang and S. H. Y. Wei, *Quintessence Int.*, 2000, **31**, 630.
214 J. W. Nicholson, *Int. J. Adhesion Adhesives*, 1998, **18**, 229.
215 C. Bentley and C. W. Drake, *J. Dent. Educ.*, 1986, **50**, 594.
216 G. A. Marynuitz and S. H. Kaplan, *J. Am. Dent. Assoc.*, 1986, **112**, 39.
217 M. Braden, B. E. Causton and R. L. Clarke, *J. Dent. Res.*, 1976, **55**, 730.
218 M. Braden and R. L. Clarke, *Biomaterials*, 1984, **5**, 369.
219 J. W. Nicholson, H. M. Anstice and J. W. McLean, *Br. Dent. J.*, 1992, **173**, 98.
220 L. H. Mair, *Quintessence Int.*, 1998, **29**, 483.
221 G. Willems, P. Lambrechts, M. Braem and G. Vanherle, *J. Dent.*, 1993, **21**, 74.
222 V. Qvist, L. Laurberg, A. Poulsen and P. T. Teglers, *J. Dent. Res.*, 1997, **76**, 1387.
223 I. A. Mjör, *Quintessence Int.*, 1996, **27**, 171.
224 N. H. F. Wilson, F. J. T. Burke and I. A. Mjör, *Quintessence Int.*, 1997, **28**, 245.
225 G. J. Mount, *Quintessence Int.*, 1997, **28**, 639.
226 W. S. Hawthorne and R. J. Smales, *Aust. Dent. J.*, 1997, **42**, 59.
227 R. L. Clarke, in *Polymeric Dental Materials*, ed. M. Braden, R. Clarke, J. Nicholson and S. Parker, Springer-Verlag, Heidelberg, 1997.
228 J. R. Lambrecht and W. L. Kydd, *J. Prosthet. Dent.*, 1982, **12**, 865.
229 D. W. Jones, E. J. Sutow, G. C. Hall, W. M. Tobin and B. S. Graham, *Dent. Mater.*, 1988, **62**, 421.
230 M. Braden and P. S. Wright, *J. Dent. Res.*, 1983, **70**, 210.
231 S. Parker, in *Polymeric Dental Materials*, ed. M. Braden, R. Clarke, J. Nicholson and S. Parker, Springer-Verlag, Heidelberg, 1997.
232 M. Nishiyama and T Katon, *J. Nihon Univ. Sch. Dent.*, 1987, **29**, 100.
233 B. I. Johansson, *J. Endodont.*, 1980, **6**, 781.
234 C. E. Friedman, J. L. Sandwick, M. A. Heuer and G. W. Repp, *J. Endondont.*, 1977, **3**, 304.
235 R. A. Augsberger and D. D. Peters, *J. Endodont.*, 1990, **16**, 492.
236 S. Limkangwalmongkol, P. V. Abbott and A. B. Sadler, *J. Endodont.*, 1992, **18**, 535.
237 *Orthodontic Materials*, ed. W. A. Brantley and T. Eliades, Georg Thiemé Verlag, Stuttgart, 2000.
238 R. Maijer and D. C. Smith, *Am. J. Orthod.*, 1982, **81**, 43.
239 J. C. Aird and P. Durning, *Br. J. Orthod.*, 1986, **14**, 192.
240 M. L. Swartz, *J. Clin. Orthod.*, 1988, **22**, 82.
241 S. D. Springate and L. J. Winchester, *Br. J. Orthod.*, 1991, **18**, 203.
242 T. Eliades, A. D. Viazis and G. Eliades, *Am. J. Orthod. Dentofac. Orthop.*, 1991, **99**, 369.
243 C. Charles, *Biomaterials*, 1998, **19**, 589.
244 T. J. Evans, M. L. Jones and R. G. Newcombe, *Am. J. Orthod. Dentofac. Orthop.*, 1998, **114**, 32.

CHAPTER 6

Biological Interactions with Materials

1 Introduction

The purpose of all biomaterials is to replace, repair or augment the natural tissue of the human body, which may be necessary as a result of disease, trauma or surgery. It follows from this that a vital aspect of the subject of biomedical materials science is an understanding of the processes that occur in response to placement of the artificial material. This can be viewed as a consideration of what takes place on the biological side of the interface between the prosthetic material and the body.

As has already been made clear, artificial materials may be implanted in a variety of locations of the body: the eye, the cardiovascular system, the hard tissue (*i.e.* the bones or teeth), and so on. The response to such implantation is a function of the location within the body and of the chemistry and physical form of the material.

At the most fundamental level, each patient is unique, in terms both of genetic make up and of their general physiology and biochemistry. However, there are sufficient similarities between people to make possible the drawing of broad general conclusions about the progress of the healing process, and how that is modified by the presence of foreign materials and devices. The response may be to react to the release of toxic components. It may be to interact with a benign surface tailored to exhibit good biocompatibility within the specific location of the body, whether bone, soft tissue or blood. It may involve the possibility of complex response mechanisms, such as blood clotting, or more simple processes, such as acceptance of bonded ceramic within the bone or a dental cement within the tooth. These topics all involve an understanding of basic biological processes, and are considered in the following sections of the current chapter.

2 Biocompatibility

Biocompatibility has been defined as 'the ability to perform with an appropriate host response in a specific application'.[1] This definition has been subject to debate, and has been described as effectively one of 'biofunctionality'.[2] Whether this is justified criticism is a matter of opinion, but what is clear is that the concept of biocompatibility has undergone a change in emphasis over many years, and is now considered to be related to those properties that enable a material or device to function satisfactorily *in vivo*, rather than being concerned with inertness. Indeed, as this chapter will show, there is considerable doubt whether any material can be truly inert within the body; even materials such as extremely passive metals, such as gold, have been shown to elicit some sort of reaction from cells *in vitro*, and would therefore be expected to do the same *in vivo*.[3]

The definition of biocompatibility has also been challenged on another front, namely that, since all interactions of biomaterials are essentially interfacial, biocompatibility needs to be thought of in these terms.[4] It is certainly true that biocompatibility of a material is largely determined by the quality of the interactions with proteins and cells, and these are of four types:

(i) Strong, non-specific attachment of cells;
(ii) Weak interaction, with no actual attachment of cells, because of lack of adhesion between them and the surface;
(iii) Strong interaction with particular sites on the surface, involving specific receptor sites on the cells;
(iv) The development of an encasement around the cells, which are then able to continue in a normal physiological manner.

However, despite the undoubted importance of these surface interactions, the definition in terms of an appropriate host response to a material placed in a specific location is the one that is most widely used, and it is useful because it highlights several important aspects of biocompatibility,[5] principally:

(i) That biocompatibility is not a single property, nor based on a single phenomenon. Rather, it refers to a collection of processes that occur as an artificial material and the tissues interact.
(ii) The definition refers specifically to the ability of the material to perform a function, reflecting the fact that the material is placed in the body for a functional reason, and has to do more than simply reside or survive *in situ*. The ability to function, especially over the long term, is heavily dependent on the response of the tissues.
(iii) The definition contains explicit recognition that inertness is probably not achievable, and circumvents this by stipulating that any response should be acceptable in view of the function to be performed.
(iv) There is explicit acknowledgement that biocompatibility *per se* is a meaningless concept, but is only applicable with reference to a specific situation in which the material is to be used. That specific situation is not only location within the body, though it includes that. It also takes account of the physical form of the material. As will be demonstrated

Table 6.1 *Adverse effects of the tissues on materials*

Material type	Possible degradation
Polymer	Depolymerisation
	Crosslinking
	Oxidation
	Hydrolysis
	Leaching of additives
	Crazing and stress cracking
Metal	Oxidation
	Crevice, pitting and/or galvanic corrosion
	Metal-ion release from passivating metals
	Stress corrosion
Ceramics	Ageing of oxide ceramics
	Dissolution

later in this chapter, a solid monolith of a material may elicit a quite different response from the same material in particulate form.

The tissues present an environment that is hostile and aggressive towards synthetic materials,[6] and the extent to which the material suffers in that environment has an influence on that material's biocompatibility. The combination within the body of an electrolytic solution containing active biological compounds, such as enzymes, as well as oxygen-based intermediates and free radicals, creates a highly reactive and damaging environment for the biomaterials and, depending on its basic nature, it can suffer a variety of modes of degradation. The most significant of these are listed in Table 6.1.

Responses to the implantation of biomaterials can be highly varied. Locally, there is usually inflammation, an important early stage in wound healing, and this may be followed by repair, often involving isolation of the artificial material with fibrous capsule.[7] If there are long-term interfacial reactions, there may be release of chemical species (monomers, low molecular weight additives, ions) into the tissues, and these may be transported around the body in either the lymphatic or vascular systems. Following this, the foreign substance may provoke a variety of responses, including hypersensitivity, carcinogenesis or accumulation in one or more of the major organs of the body. These are discussed in the relevant sections of the current chapter.

Given the complex range of possible adverse reactions to artificial materials, the testing of candidate materials for biocompatibility is critical. Many of the current methods are cell-based *in vitro* tests, aimed at determining very specific effects on cells, rather than animal-based, and aimed at considering systemic effects. Nonetheless, phenomena in the latter category remain important, and are themselves the subject of ongoing research.

Tests for biocompatibility vary from the simple, based on straightforward observations of cell viability in cultures, to the more complex, using the methods of cell and molecular biology. There continues to be pressure to reduce the volume of animal-based tests and, though these cannot be eliminated completely, much of the current emphasis, including the current ISO standards, is on cell culture and related methods, rather than on experiments involving whole animals.

The following test methods are available and are all quantitative in character, as well as in reasonably widespread use:

(i) The MTT test. This test measures cell metabolic function and is dependent on the intact activity of succinate dehydrogenase, a mitachondrial enzyme, whose function becomes impaired if the cells are exposed to a toxic substance. The test employs the salt 3-(4,5-dimethylthiazol-2-yl)-2,5-diphenyltetrazolium bromide, MTT, and involves its conversion to an insoluble formazan product which can be measured spectrophotometrically.[8] It has been widely used, for example in the testing of orthopaedic biomaterials.[9] In this study, human osteosarcoma cells were used, and the high sensitivity of the MTT assay was demonstrated.

(ii) DNA synthesis. As cells develop and, in cell cultures before they have become confluent, they synthesise DNA immediately prior to undergoing mitosis. Cell cultures at this stage can be pulse-labelled with a nucleotide of DNA which has a tagged atom, typically a radio-isotope such as tritium, included in it to enable detection. Alternatively, a functionally substituted molecule, such as bromodeoxyuridine, BrdU, can be used, with detection by means of a fluorescent dye-conjugated monoclonal antibody once the BrdU has been taken up.[10] This test is based on the fact that rapidly proliferating cells will take up more of the tagged component into them than cells which are proliferating either slowly or not at all.

(iii) Membrane integrity tests. If cells are exposed to toxic substances, their membranes undergo changes that lead to alterations in function. This can be tested by incubating the cells with a mixture of fluoroscein diacetate and ethidium bromide. The fluoroscein diacetate is taken up by intact cells and converted within the cell to fluoroscein, which gives a green fluorescence under light of the appropriate wavelength. By contrast, the ethidium bromide cannot enter the cell when the membrane is intact. However, if the membrane becomes damaged, it passes into the cell, where it binds strongly to nucleic acids, especially DNA. The resulting complexes give an orange-to-red fluorescence under the appropriate wavelength.[11]

Isolated cells can be very sensitive, and it might be that they become more sensitive to toxic effects of biomaterials than they would be within the body. Consequently, an alternative approach to *in vitro* testing is to use organ cultures. For example aortas from chicks have been used in the evaluation of vascular prosthesis materials[12] and explant cultures of human gingiva have been used to evaluate dental materials.[13] This approach is able to give more realistic results in certain circumstances, especially where some artificial surfaces are unable to support the growth of individual cells because of surface properties rather than cytotoxicity.

3 Toxicity

The placement of a foreign material in the body carries with it the possibility of toxicity, especially if low molecular weight fragments, such as metal ions or oligomers, can be lost from the implant. The subject of toxicology is a large one, and its more specific aspects will be considered in this section.

In toxicology, as in pharmacology, the receptor theory is the basis of understanding.[14] This theory is supported by the fact that specific receptor proteins are known which are associated with pharmacological interactions. It seems likely that similar substances are available which initiate a toxic reaction, and these targets may be not only proteins, but also nucleic acids, enzymes and membrane components.

Toxicity may be acute or chronic, and these two possibilities are distinguished by operational criteria and by underlying mechanisms at the molecular level. Acute toxicity is that which arises soon after administration, and results in either death or complete recovery. Chronic toxicity, by contrast, may not be apparent immediately, but take time to develop, and may not result in a clear cut consequence, but rather in prolonged ill-health. Chronic toxicity may be divided into two types:

  (i) that which requires prolonged or repeated administration;
  (ii) that which requires one (or very few doses) but which has a long lasting effect.

Clearly type (i) chronic toxicity is a possibility with biomaterials. Gradual release of an appropriate chemical species, for example metal ions, as a result of gradual corrosion of an implant *in vivo* has the same effect as prolonged or repeated administration of a toxic substance.

It is usually the case that a particular organ is affected by the toxic agent. This is known as the target organ for the poison in question. An example of a highly toxic substance is cadmium, and this has its target organs the liver and the kidneys. Continuous exposure to small amounts of cadmium by the patient leads to the development of much larger concentrations within these organs. This occurs because cadmium forms stable complexes with metallothionein, which is present in both the liver and the kidneys. When complexed in this way, the cadmium is much less mobile than in the free state, and hence it accumulates in these organs. Once the concentration has built up to $300–400\ \mu g\ g^{-1}$, necrosis occurs.[15] This mechanism of toxicity has been established partly because cadmium-induced necrosis of the kidney is associated with excretion of the protein complex cadmium thionein in the urine. Similar effects in terms of accumulation at key organs occur with metals such as vanadium and aluminium, both of which find use as alloying metals with titanium for implants in total hip and total knee replacement surgery. Vanadium accumulates in the lung tissue, while aluminium becomes concentrated in the surrounding muscles, lungs and regional lymph nodes.[16] Unlike cadmium, these metals do not seem to be associated with necrosis, or other adverse effects.

4 Cytotoxicity

Loss of viability by cells due to their interaction with artificial materials can be used as the basis of determining the likely biological effects of these materials *in vitro*. Tests of this type may be used to screen innovative materials[17] prior to animal tests, and they may also be used for quality control of both raw materials and finished devices.

Cytotoxicity has been defined as the *in vitro* evaluation of toxicological risks using cell culture. The tests can give information on cell viability and death at the grossest level, to details and sophisticated findings concerning the state of the cell membranes, cell organelles, protein or DNA synthesis, and cell division.

Test methods to evaluate cytotoxicity are of the following types:

(i)   contact with liquid extracts prepared from the material;
(ii) direct contact with the material;
(iii) indirect contact (diffusion) using either liquid extracts or the materials themselves.

An example of the latter test method that is widely used is the agar overlay test. In this test, a sample of the material to be evaluated is placed on a solidified agar-culture medium mixture covering a near confluent monolayer of cells of a selected cell line. Cytotoxicity is determined, usually in triplicate, following a 24 hour incubation period. A variety of cell lines can be used, but clearly it is most relevant to use cultured human cells of some sort, such as HeLa cells.[18] However, other mammalian cells, such as mouse fibroblasts are also used.

A difficulty with cytotoxicity testing is that the cell culture is a closed, non-buffered system, and this makes it significantly different from the potential site of implantation of the biomaterial. Consequently, results from cytotoxicity studies have to be interpreted with care. For example, degradable polymers such as polylactides show a significant level of cytotoxicity in cell culture tests, yet these materials are acceptable in the body, are safe and show little or no adverse effects on cells *in vivo*. The reason for this discrepancy is that, in the cell culture, degradation products, mainly lactic acid monomer, cannot escape and are not buffered. The cells therefore experience a catastrophic drop in pH and consequently die. In the body, by contrast, the lactic acid degradation product is buffered, thus preventing any drop in pH. It is also removed from the site of implantation, and therefore unable to build up to toxic levels.

Cytotoxicity tests are concerned with monitoring cell death, but it is important that the cause of cell death be established and, in particular, that the possibility of apoptosis having occurred be taken into account.

Apoptosis, or programmed cell death, is a natural process that takes place under physiological conditions.[19,20] Its features are:

(i)   that it is necessary under certain physiological conditions, *e.g.* embryo-genesis, homeostasis of the immune system, *etc.*;
(ii) that is a major mechanism for regulation of cell numbers;
(iii) that it is necessary as a means of removing self-active lymphocytes and tumour cells.

Once initiated, apoptosis begins with a pre-commitment stage during which a high level of secondary messenger substances, such as Ca^{2+}, IP_3 and/or cAMP build up in the cell. This is followed by a series of biochemical changes, as follows:

(a) DNA is caused to break down by the action of an endonuclease, to yield ultimately very small oligonucleosomal fragments.
(b) Towards the end of the process, the plasma membrane undergoes lysis, and the cell fragments that result are phagocytosed either by macro-phages or by neighbouring cells.

The process is genetically controlled and is initiated either internally or by external agents, such as hormones, cytokines, killer cells or a variety of chemical, physical or viral agents. It can occur very quickly, and be complete in a matter of minutes, which makes it very difficult to observe in tissue sections.

The biochemistry of apoptosis is becoming better understood, and many of the substances involved in it have been identified and their roles elucidated. For example, the cytokines TNF-α and IL-1β are known to be involved.[21,22] Of these, TNF-α is one of the important extracellular activators of apoptosis in the mammalian immune system. IL-1β is derived from IL-1β converting enzyme, ICE, an enzyme that plays a critical role in inducing apoptosis. Activated ICE acts on non-functional pro-IL-1β, converting it to IL-1β, which is the active form for apoptosis.

Apoptosis is regulated by the bcl-2 family of proteins.[23,24] They either promote or suppress apoptosis by interacting or antagonising each other.[25] In the presence of excess bcl-2, heterodimers of bcl-2 and bax are formed, and these prevent apoptosis. On the other hand, where bax dissociates from bcl-2, or in an excess of bax, homodimers predominate, and these promote apoptosis.

Apoptosis can be distinguished from so-called oncosis, formerly more widely known as necrosis, or accidental cell death by differences in the morphology of the cells. These are detailed in Table 6.2.

Several genes appear to be involved in the process of apoptosis, and when their activity is suppressed, there are pathological consequences. Certain anti-cancer drugs are known to be capable of inducing apoptosis, so that its occurrence is not fixed, but related to the chemistry of the environment in which the cell finds itself. This has consequences for cell–material interactions.

Table 6.2 *Morphological features of cells undergoing apoptosis or oncosis*

Apoptosis	Oncosis
Cell shrinks	Cell swells up
Cell becomes rounded	Protein denaturation and hydrolysis
Chromatin becomes pyknotic	Cell contents released
Nucleus breaks up with cell budding	Local tissue damage
Little of no inflammation of mitochondria	Intense inflammatory response

Table 6.3 *Metals known to induce hypersensitivity*

Metal	Reference	Comment
Be	Liden *et al.*[28]	
Ni	Haudrechy *et al.*[29]	Commonest cause of metal sensitisation in humans
Co	Liden *et al.*[28]	
Cr	Yang and Merritt[30]	
Ta	Angle[31]	Ta, Ti, V show occasional responses only
Ti	Lalor *et al.*,[32] Parker *et al.*[33]	
V	Angle[31]	

However, it is still not clear how apoptosis can be induced, nor whether it is prevalent where artificial materials are known to show cytotoxicty. More work is needed to elucidate the mechanism of apoptosis, and also to determine how the presence of foreign materials, such as wear particles from artificial hip or knee joints, might bring apoptosis about.

5 Hypersensitivity

When materials corrode in a biological system, they release ions. These ions are then able to form complexes with proteins, and the resulting metal–protein complex may act as an antigen (or allergen) in the patient.[26] Metal hypersensitisation is well established[27] and its appearance following contact of the skin with metals, corrosion products or metal salts has been reported as affecting some 10–15% of the population.

Metals vary in their capability for inducing hypersensitivity, and the main ones causing such effects are listed in Table 6.3.

Cross-sensitivity reactions between metals may occur, most notably between nickel and cobalt. What this means is that the initial metal–protein antigen confers upon the body the ability to raise antibodies that respond not only to the original metal–protein complex but also to the complex of the cross-sensitive metal

Dermal hypersensitivity typically presents in the form of skin lesions of various types: hives, eczema, redness and itchiness. Within the body, the generation of degradation products from biomaterials, either ionic corrosion products or particulate wear debris, gives rise to immunological reactions. Such adverse reactions have been reported for a variety of metallic devices, including those for cardiovascular,[34] orthopaedic,[35,36] and dental[37] applications. Metals involved have mainly been nickel,[38] cobalt and chromium, the latter arising either from cobalt–chromium alloys[39] or from stainless steel.[40]

Hypersensitivity can be an immediate humoral response that is apparent within minutes, or a delayed response that is not apparent until several hours or even days have elapsed. The immediate response is due to the formation of antibody–antigen complexes, and these are divided into Type I, II, or III reactions. By contrast, the delayed response, so-called Type IV, is a cell mediated reaction. These responses are now considered in detail:

(i) Type I response. This is typified by the binding of soluble allergens (antigens) to B lymphocytes, and these then transform to IgE-secreting plasma cells and memory cells.[41] IgE binds to Fc receptors on basophils or mast cells and this has the effect of sensitising them. When the basophils or mast cells are next exposed to the sensitising allergen, there is a reaction that releases pharmacological substances that promote vasodilation, increased vascular permeability and smooth muscle contraction. The resulting manifestations include systematic or localised anaphalaxis, hay fever, asthma, hives and eczema. Allergens for such a response include plant pollens, drugs, foods (notably nuts) and insect venoms. The typical initiation time of Type I responses is 2–20 minutes.

(ii) Type II response. By contrast with the Type I response, this is mediated by antibodies rather than IgE, and is characterised by activation of the complement system or cytotoxic T cells, and these eliminate any cells that display the antigen. Host antibodies reacting with foreign antigens produce spores in the membrane of foreign cells or serve as targets for guiding phagocytic cells. Typical antigens include transfused blood proteins and maternal IgG antibodies which can cross the placenta and destroy foetal red blood cells. The typical initiation time of a Type II response is 5–8 hours.

(iii) Type III response. This response is immune-complex mediated, and involves large amounts of circulating antibody that is specific to an invading antigen. These form antibody–antigen complexes that build up to give high local concentrations, and these high concentrations cause increased vascular permeability and stimulate chemotactically active neutrophils. This leads to local accumulation of fluids and red blood cells. If the response is mild, the signs are redness and swelling; if it is severe, there may be tissue necrosis. Typical antigens include insect bites, bacterial spores and, most commonly, antitoxins such as anti-tetanus and anti-diphtheria serum. Initiation time for a Type III response is in the range 2–8 hours.

(iv) Type IV response. In this response, activated and sensitised T-DTH lymphocytes release various cytokines, and these result in the accumulation

and activation of macrophages. In a fully developed Type IV response, only some 5% of the participating cells are antigen-specific. The primary effector cells and most of the other participating cells are macrophages. Hypersensitivity reactions provoked by implants tend to be Type IV responses and a typical initiation time is 1–3 days.

Metal hypersensitivity was first shown to be associated with metallic orthopaedic implants in the mid 1960s,[42] and this has since been confirmed in a number of other studies.[43–45] Reactions of affected patients have been found to range from mild skin rash to weeping blisters. The principal metal causing these problems is nickel, as was established by patch testing using nickel sulfate solution. Typically these skin problems have been found to abate once the implant has been removed.[46,47]

Establishing the occurrence of Type IV responses is not straighforward. Patch testing has been widely used, typically by incorporating the suspect antigen, such as 1% aqueous nickel sulfate, in a carrier such as petrolatum, and exposing a small area of the skin to this by means of a bandage. After an exposure time of 2–4 days, the skin is examined and any reactions graded on a standardised scale running from 1 (mild or no response) to 4 (severe redness, with possibly small weeping blisters). The difficulty lies in the fact that this is a substantially different exposure regime from the long-term constant exposure that implants provide, and which can trigger eczemic reactions. The potential of metal ions to trigger an effect through dermal contact would be expected to be different from that of metal ions released systemically from an implant. There is also the possibility that such testing may in exceptional cases induce hypersensitivity in a previously unaffected patient.[48]

6 Carcinogenicity

Cancer is the name given to malignant tumours, that is, to tumours which have the capacity to invade neighbouring tissues and to spread to remote sites of the body through the blood or lymphatic systems. At the remote sites, subpopulations of malignant cells take up residence, grow and divide, and again invade neighbouring tissue to form secondary tumours. These are referred to as metastases. It is also possible for tumours to be benign. These do not invade neighbouring tissues, nor colonise distant parts of the body. Both types are referred to as neoplasia, a word which literally means 'new growth', and both involve excessive and uncontrolled cell proliferation.[49] The characteristics of tumours are shown in Table 6.4.

Cancer cells more or less resemble the cells from which they were derived, though malignant cells are generally less differentiated than normal cells. However, they usually possess sufficient structural similarity with the original cells that the organ of origin and the cell type can be identified. In fact, the extent to which the tumour cells resemble the original cells determines the biological behaviour, which in turn determines the expected outcome for the patient, the so-called prognosis. Poorly differentiated tumours tend to display more malignant behaviour, that is, are more aggressive, growing faster and

Table 6.4 *Characteristics of tumours*

Feature	Benign	Malignant
Differentiation	Well defined. Similar to tissue of origin	Poorly defined. May contain bizarre (anaplastic) cells
Growth	Slow and progressive. May regress	Erratic and may range in speed from slow to fast
Invasion	Generally localised and cohesive	Invasive, spreading to adjacent tissues
Metastasis	Absent	Usually present. Large, less differentiated tumours are more likely to metastasise

spreading more rapidly than those which mimic more closely normal cells from the tissue of origin.

Most cancers do not have an identifiable cause, and the overall mechanisms of carcinogenesis have not yet been determined. However, in certain circumstances, concerns have been expressed that the use of biomaterials may be linked with the development of particular types of tumour. It must be said against this, though, that, although the use of implants in surgery is growing rapidly, there has been not been a corresponding increase in reported cancers, and in particular it is extremely rare to find neoplasms at the actual site of an implanted device.[50]

On the other hand, there have been examples of tumours being reported to have developed at sites remote from the implant, but which have been associated with the presence of that implant within the body. Examples are drawn from a wide range of implant materials, including polymers and metals, and from a range of different surgical procedures. These are shown in Table 6.5.

It is necessary to exercise caution when attributing the formation of a neoplasm to the presence of an implant. Just because a tumour develops adjacent to an implant does not mean that the implant caused the tumour. Neoplasms are quite common in humans, and so can occur anyway in the vicinity of an implant.

On the other hand, foreign bodies are known to be capable of inducing tumour formation. This phenomenon seems to arise because of the physical effects of the foreign body, rather than because of its chemical characteristics.[60] Tumours have been induced experimentally by a wide range of materials, including those that are apparently almost inert, such as certain glasses, gold or platinum. Physical structures with low specific surface areas, *i.e.* monolithic pieces, present the greatest risk of developing tumours. When they are in powdered, shredded or woven form, and their specific surface area is much greater, there is much less likelihood of tumours developing. This is called the Oppenheimer effect, after the researchers who originally reported it.[61] It demonstrates that, provided the chemical reactivity is sufficiently low, it is the physical state that determines whether or not neoplasms develop.

Table 6.5 *Examples of tumours associated with the presence of implants*

Device	Materials implicated	Time of implantation (years)	Tumour	Reference
Fracture fixation	Vitalium	17	Lymphoma	McDonald[51]
		9	Osteosarcoma	Ward *et al.*[52]
Total hip	PMMA/ UHMWPE	2	Malignant fibrous histiocytoma	Bago-Granell *et al.*[53]
	Alumina	1+	Soft tissue sarcoma	Ryu *et al.*[54]
	UHMWPE	10	Osteosarcoma	Martin *et al.*[55]
	Stainless steel/ PMMA	12	Synovial sarcoma	Lamovec *et al.*[56]
Total knee	–	Not reported	Epithelial sarcoma	Weber[57]
Vascular graft	Dacron	1+	Malignant fibrous histiocytoma	Weinberg *et al.*[58]
	Dacron	12	Angiosarcoma	Fehrenbacker *et al.*[59]

Tumours of this type arise following the formation of apparently bland fibrous capsule around the implant. There is an inverse correlation with chronic inflammation. Lack of chronic inflammation seems more likely to lead to a tumour being formed, whereas an active and persistent inflammatory response appears to inhibit tumour formation. These observations have been drawn from studies of animal models, especially rats. However, rats are much more susceptible to tumour formation in response to foreign bodies than humans. Consequently there is doubt about the value of such studies of foreign-body tumours in rats and their predictive power in identifying potential clinical difficulties in human patients.

Certain foreign materials can induce tumours as a result of their chemistry. For example, chromium in its hexavalent state is a known carcinogen and the possibility exists of Cr^{VI} arising as a corrosion product from cobalt–chromium alloys used as implants. There have been clinical cases in which osteosarcomas have been found in patients who have had hip implants made of cobalt–chromium alloy, and these have been attributed to the presence of these prostheses.[62-64]

However, in general the main factors that influence the growth of cancers in response to a foreign body are implant shape, fibrous capsule formation and the amount of time available for the neoplasm to develop. With all three factors, the major effect of the implant seems to be that of stimulating cell maturation and proliferation. Despite the concerns with remote site tumours from cobalt–chromium, it is uncommon to find neoplasms associated with implants, and even rarer to find such growths associated with a localised foreign body response, at least in humans.

Table 6.6 *Sequence of events following implantation*

Injury
Acute inflammation
Chronic inflammation
Granulation tissue formation
Foreign body reaction
Fibrosis

7 Interaction of Materials with Soft Tissues

When a body is injured, healing takes place in a series of well ordered stages. Placing a biomaterial in the body is, from the biological point of view, an injury, and these typical responses need to be considered in terms of how they are modified by the presence of the artificial material. This sequence is shown in Table 6.6.

Inflammation

Inflammation is the reaction of living vascularised tissue to local damage.[65] It serves many functions, including the isolation, dilution or neutralisation of the process or agent causing the injury. Inflammation initiates a series of events that may lead to healing. In particular, it may lead to replacement of the damaged tissue by regeneration of neighbouring parenchymal cells, or to the formation of scar tissue, so-called fibrosis. Alternatively, a combination of these processes may occur.

Immediately the damage has occurred blood flow and vascular permeability are altered. This causes fluid that is rich in proteins and blood cells to escape from the adjacent blood vessels in a process known as exudation. This triggers a series of cellular events that characterise the inflammatory response. Among them is blood clotting and possibly thrombosis formation. Blood interactions are considered later in this chapter.

Following injury, there is a progressive change in the cells which predominate at the site. Initially, neutrophils are present. These are attracted to the site early on as a result of chemotactic factors released soon after the initial injury. They are short-lived, and they disappear or disintegrate within some 24–48 hours. They are then replaced by monocytes. At the site of the injury, monocytes undergo a differentiation process to become macrophages, and these cells may live for several months. Emigration of the undifferentiated monocytes towards the site of the injury may continue for days or even weeks, depending on the extent of the damage, partly because chemotactic factors for monocytes are activated over relatively long periods of time.

Although triggered by the injury, *i.e.* in this case the surgical procedure to place the implant, the nature of the biomaterial itself has an influence on the progress of the inflammatory response. Depending on its geometry, surface chemistry and mechanical properties, the implant can cause variations in the

Table 6.7 *Typical chemical mediators of inflammation*

Mediators	Examples
Vasoactive amines	Histamines, serotonin
Plasma proteases	
Kinin system	
Complement system	
Coagulation/fibrinolytic system	
Arachidonic acid metabolites	
Prostaglandins	GGI_2, TxA_2
Leukotrienes	HETE, leukotriene B_4
Lyposomal proteases	Collagenase, elastase
Oxygen-derived free radicals	H_2O_2, superoxide anion
Platelet activating factors	Cell membrane lipids
Cytokines	
Growth factors	Fibroblast growth factor

intensity and duration of the inflammatory response. Indeed, the extent to which it does so has a major influence on its biocompatibility.

Once the inflammatory response has occurred following injury, the chemicals released from the plasma, cells and damaged tissue mediate a response. These responses may vary, and they may also be aimed at destroying the implant, which is recognised as a foreign body. They may thus contribute to the degradation of the biomaterial. Some of the important chemical mediators are listed in Table 6.7.

The initial inflammation, so-called acute inflammation, is relatively short-lived, and may last from a few minutes to a few days, depending on the severity of the injury. It is characterised by exudation of fluid and plasma proteins, and also by migration of leucocytes, mainly neutrophils.[66] These latter are white cells from the blood that move to the site of the injury, their principal function being to engulf invading micro-organisms and foreign materials in a process known as phagocytosis.

Biomaterials are rarely small enough to be phagocytosed by either neutrophils or macrophages. However, some of the steps of phagocytosis may occur, including attachment of the cells to the biomaterial surface and the release of leukocyte products aimed at degrading the implant. The amount of such material released seems to vary with the dimensions of the implant, and there is evidence of greater release when the material is in powdered form, and hence potentially able to be phagocytosed, than when the material is present in bulk.[67]

If the inflammatory stimuli remain, acute inflammation may give way to chronic inflammation. This varies more than acute inflammation, and the cells present at the site may differ. The longer-term inflammatory stimuli that provoke this chronic response may include motion of the implant within the site of placement, and sustained release of damaging chemical species, such as corrosion products from metals.

The cells present at the site in chromic inflammation are mononuclear, and include lymphocytes and plasma cells. Such cells are often associated with immune reactions, though their roles in this type of response to implant materials has not been elucidated. Macrophages may also be present in significant numbers, and these are known to be capable of a range of responses, and the release of a variety of chemicals. These include neutral proteases, chemostatic factors, arachidonic acid metabolites, growth factors and cytokines.[68]

Granulation Tissue Formation

Granulation tissue is so-called because of its granular appearance. It is soft, pink tissue which develops on the surface of healing wounds, and it includes numerous small blood vessels and fibroblasts. Such tissue may be observed within a few days of the implantation procedure.

The blood vessels are formed within the granulation tissue in large numbers by a process of budding of existing vessels. This involves proliferation, maturation and organisation of the adjacent endothelial cells to form capillaries. Within the granulation tissue there are also numerous fibroblasts, and these proliferate and function to generate collagen, mainly type III, and proteoglycans. In addition, macrophages are usually present.

The extent to which wound healing can take place depends on the precise details of the implantation operation. If a considerable degree of injury has been necessary, or a large defect created, wound healing will be less extensive and complete than if the procedure has been much less intrusive. The biological function of granulation tissue is full reconstruction of the original form of the body. However, where this is not possible, and large amounts of granulation tissue are formed, the result may be fibrosis or the formation of scar tissue. First, however, we need to consider the possible reaction of the body to the inclusion of the implant, a material which is by definition a foreign body.

Foreign Body Reaction

When a biomaterial provokes a foreign body reaction, this involves so-called foreign body giant cells and the components of granulation tissue, *i.e.* macrophages, fibroblasts and capillaries. The details of the foreign body reaction vary with the form and surface of the implant. Flat, smooth surfaces have a foreign body reaction that consists of a layer of macrophages one or two cells thick. Rough surfaces, such as those on the outer parts of PTFE vascular prostheses, have a foreign body reaction that consists of macrophages and foreign body giant cells at the surface.

The cells deposited as the foreign body reaction may remain for the lifetime of the implant. Typically, an implant becomes surrounded by fibrous capsule, and this develops around the foreign body reaction as well, thereby isolating both it and the implant. Early on in the process of inflammation and wound healing macrophages are activated upon adhering to the implant surface. What influences the activity subsequent to this is not clear. Certainly, foreign body giant cells have been observed on biomaterial surfaces in numerous studies, and they are known to be capable of persisting for the lifetime of the implant but, whether they continue to release lysosomal constituents, or whether they become deactivated, is not known.

Fibrosis

The formation of fibrous capsule is often the final part of healing following implantation of a biomaterial. This is not always the case and materials such as titanium alloys implanted into bone will form a close union with the regenerated bone that does not involve fibrous capsule. However, in the soft tissues, this is a much more frequent consequence of implantation.

Essentially, fibrous capsule develops when adjacent parenchymal cells are replaced by connective tissue. This occurs because the cells involved are typically permanent, *e.g.* nerve cells, skeletal or cardiac muscle cells. Unlike so-called labile cells, such as those found in the lymphatic and blood systems, these cells are unable to reproduce themselves after birth. Hence, the only repair mode open to them is fibrosis.

8 Interaction with Blood

When blood vessels are damaged, blood is lost to the surroundings, a process that in healthy subjects does not continue for very long, as the blood clots and the body begins the process of repair. Blood clots readily as a result of a complex biochemical process and, though highly desirable under most circumstances, it can cause problems where biomaterials are concerned. Blood can be induced to form clots (thromboses) under a variety of circumstances by biomaterials:[69] if it becomes stagnant in localised regions around prosthetic heart valves, if it flows over low energy surfaces, or if it flows over high energy surfaces.[70] The topic of blood clotting is central to an understanding of the topic of haemocompatibility[71] and will be considered in outline in the following paragraphs.

A key component of the blood that assists coagulation is the platelet. Platelets have two functions, namely (i) to arrest initial bleeding through the formation of plugs and (ii) to stabilise the platelet plugs by catalysing the clotting reactions, leading to the formation of fibrin. The platelets themselves are disk-shaped cells about 3.5 μm in diameter, but without nuclei. They are produced in the bone marrow and circulate in the blood where they occupy an estimated 0.3% of the total blood volume.

Under normal circumstances, platelets retain their disk-like morphology, but when activated they become irregularly shaped, with pseudopods arising from an essentially spherical core. They also become sticky. These changes allow the platelets to bind together to form aggregates that are the basis of thrombi or blood clots. In this state they are also able to adhere to the walls of injured blood vessels, and also to the surfaces of artificial materials placed as implants.

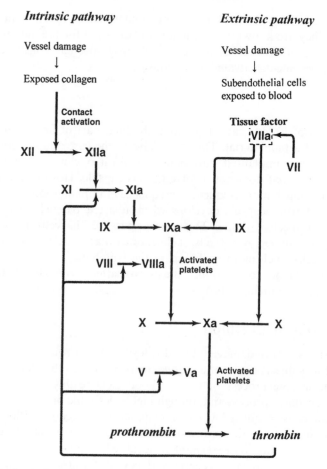

Figure 6.1 *Intrinsic and extrinsic coagulation pathways of blood*

The actual process of blood clotting is extremely complex. At least twelve plasma proteins are known to be involved, and they interact in a series of complex reactions. These proteins are designated by Roman numerals, which were applied in the order in which the proteins were discovered, which usually occurred before their role on the coagulation process was determined. Two systems have been identified, the intrinsic system and the extrinsic system, and these are illustrated in Figure 6.1.

Other components of blood include red and white cells. Of these, the red cells are usually considered the passive participants in the processes of haemostasis and thrombosis. However, under certain conditions, such as low shear rates or venous flows, they may contribute significantly to the mass of thrombus formed. Red blood cells, through the effects of their concentration and motion, have an important mechanical influence on the diffusion of blood elements. Under certain conditions, too, they may release factors that influence platelet activity.[72] They will attach to artificial surfaces, though this is not important in the overall promotion of biocompatibility, and has not been studied particularly closely.

White blood cells are of various types, and they perform a range of important functions in inflammation, infection, wound healing and the response of the blood to foreign materials. White blood cells therefore have an interaction with biomaterial surfaces, though the details of this are still far from clear. White blood cell interactions with implant devices with large surface areas may be considerable, and may explain the observation that these cells become depleted in the circulating blood when such implants are present.

Red and white blood cells circulate in suspension in the blood plasma. Also present in the plasma are a variety of components which are required for antibody-mediated bactericidal activity. This is termed the complement system, and is a complicated and finely balanced biochemical system composed of more than 20 distinct plasma proteins. Two distinct pathways are known through which the complement system functions, the so-called classical and alternative pathways. The complement system is a contributor to the acute inflammatory reaction, and one of its principal functions is recognition and elimination of foreign elements from the body.

Both pathways for complement activation involve the component C3. The classical pathway is initiated by the C1q subunit binding to the Fc portion of an antigen–antibody complex.[73] This induces a conformational change in the C1 complex, resulting in a number of structural changes: C1r and C1s dissociate from C1q; C1s than cleaves C4 into C4a and C4b. The C4b binds C2, and C1s cleaves C2. The C4b2b complex acts as a convertase for C3, promoting cleavage into C3a and C3b.

The alternative pathway begins with cleavage of C3. Activation then continues along the terminal pathway, with activation mediated or regulated by Factors B, D, H and I, and by properdin. Most biomaterials seem to activate the complement system *via* the alternative pathway,[74] and an important aspect of this activation is the adsorption of $C3.H_2O$ onto the surface. This is followed rapidly by hydrolysis of the C3, a process which leads to a conformational change that exposes the thioester group of the C3b portion.[75] The polarity of this functional group allows the C3b portion to bind to a variety of nucleophilic reagents, such as water, hydroxyl groups and amines, and this can lead to a variety of subsequent reactions.

Hydroxyl groups on the surface of materials are critical to the promotion of complement activation, and a linear relationship has been demonstrated *in vitro* between the hydroxyl content of acrylic polymers and the generation C3a in plasma.[76] Related to this is the observation that complement activation by regenerated cellulose dialysis membranes can be decreased by forming derivatives of hydroxyl groups, for example, by converting them to acetate groups.[77] This approach results in a membrane with fewer available binding sites for C3b on the surface, and thus inhibits the formation of the convertases necessary for activation to continue.

In the body, a number of possible interactions have been found to occur

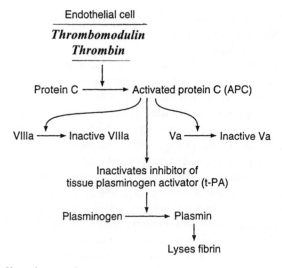

Figure 6.2   *The fibrinolytic pathway*

between the complement system, coagulation of the blood and fibrinolysis. Consequently, there has been much interest in the fact that artificial surfaces can activate the complement system. These have been prompted in part by the observation that devices having large surface areas, such as haemodialysers, are capable of causing reciprocal activation reactions between complement enzymes and white blood cells.[78] When such artificial surfaces activate the complement system, both white blood cells and platelets are modified and become adherent. This may have significant clinical consequences. For example, following haemodialysis, the increased adherence of the white blood cells has been found to cause them to become trapped at the lung surfaces, while at the same time becoming depleted in the peripheral regions of the body.[79] This condition is known as neutropenia, and has been shown to be directly correlated with extent of complement activation with a variety of dialysis membranes.

Another function of the blood is fibrinolysis, that is, the removal of unwanted fibrin deposits following thrombus formation, thereby improving blood flow. This process, which is mediated by a multi-component system in the plasma, also facilitates the healing process following injury and inflammation. As with the other systems, the fibrinolytic system is complex, and consists of precursors, activators, cofactors and inhibitors. It also interacts with the coagulation system at the level of contact activation. A simplified version of the fibrinolytic pathway is shown in Figure 6.2.

The best understood process involves the fibrinolytic enzyme plasmin. Under normal circumstances, this circulates in the blood in an inactive form known as plasminogen. This protein will adhere to a fibrin clot, and becomes gradually incorporated into the mesh. It becomes activated to plasmin by the action of specific activators. These may be present in the blood, released from the tissues, or administered as a drug. Once activated, the plasmin degrades the fibrin clot, releasing soluble digestion products into the circulating blood.

Overall, then, the current state of knowledge is that there are numerous related biochemical systems within the blood and these respond rapidly to tissue injury. This rapid response minimises loss of blood, but also requires the removal of excess deposits once healing is complete. These are all achieved satisfactorily in a normal healthy individual. Unfortunately, artificial surfaces can destroy the delicate balance between the activation and inhibition parts of these systems, and this may lead to excessive thrombus formation or to an exaggerated inflammatory response. These adverse effects mean that the materials and devices that can be used in repair or replacement within the cardiovascular system are limited. There is currently considerable interest in understanding how the clotting of blood is triggered by contact with artificial surfaces, and also in the search for new and more versatile haemocompatible materials.

Various reactions occur when blood comes into contact with a foreign material, and these may be influenced by a number of factors related to the material surface including chemical composition, surface energy, roughness and topography.[80] The events that occur following implantation of a material have been suggested to be as follows:[81]

1. Interaction with water molecules. These arrive at the implant surface and bond to it *via* hydrogen bonds involving either the hydrogen or oxygen atoms of the water. This may be followed by adsorption of a second water layer that may also include small ions.

2. Interaction with proteins. This is a complex set of interactions, which is discussed in more detail in the following section. It begins with an interaction of proteins with the initially adsorbed later of water molecules. The details of this process are dictated by the nature and orientation of the initially bound water molecules, and determine in turn how subsequent interactions proceed. The most common proteins which can be deposited on the surface are fibrinogen, γ-globulin and albumin, but other proteins may be included in smaller quantities. These include fibronectin, transferrin, von Willebrand's factor and the coagulation factors XI and XII.

These surface/protein interactions determine the haemocompatibility of the surface. Surfaces that preferentially adsorb albumin rather than fibrinogen show good haemocompatibility. They are resistant to thrombosis formation, due to the effect that albumin has in minimising platelet adhesion. Conversely, surfaces that favour the adsorption of fibrinogen and γ-globulin are much less haemocompatible, because of the ease with which thrombosis formation can occur, with associated platelet interaction and activation. The presence of γ-globulin on the surface particularly enhances both platelet and granulocyte adhesion, and also stimulates platelet release.

Following the adsorption of plasma proteins onto the surface, development of the crosslinked fibrin network may take place. The timescale for this may vary considerably, from a few minutes to several hours, but the longer it takes, the less the chance of gross thrombus formation.

Interactions of this type with an artificial surface can cause one of the plasma proteins, Factor XII, of the intrinsic coagulation pathway to be initiated, either by contact with the artificial surface or by contact with the adsorbed aggregates of platelets. Other haemostatic reactions occur that involve platelet activation, the complement system and tissue factors. All of these coagulation processes are interconnected, and must be considered in assessing the haemocompatibility of a material.[82]

Which coagulation pathway is most important varies with the location of the implanted material. High shear stress environments are more likely to activate platelets, while low shear stress environments are more likely to trigger protein-based coagulation processes.[83,84] A prosthetic heart valve is in a location where both low and high shear stresses can occur, so that both platelet activated and protein-based coagulation processes may take place. Appropriate design of the valve can improve the balance of shear stresses around the implant, but the material of construction is also critical in determining the haemocompatibility of the device.

The response of platelets to implant materials has been extensively studied. Platelets may adhere to the foreign surface[85] following which they may or may not activate.[86] Such adhesion may be followed be spreading, to create a passivating layer over the surface, though this does not always occur. Alternatively, the platelets may activate despite not adhering to the surface.[87]

The platelet response, in the short term, has been suggested to be the most significant factor in determining the haemocompatibility of a material.[88] On the other hand, the effects of other components may be more significant for the longer-term survival and acceptability of blood-contacting devices. Of these, the alternative pathway of the complement activating system is important, because it allows for non-specific activation by foreign materials. This is acheived through direct activation with C3, which is converted to C3a anaphylotoxin. Complement activation of this type has been shown to occur as a result of using extra-corporeal circuits in cardiopulmonary bypass, as well as with haemodialysis membranes, and such complement activation leads to increased leukocyte adhesion to activating surfaces as well as to increased likelihood of thrombosis.[89]

A number of polymeric biomaterials have been shown to activate the complement system *via* their influence on Factor XII. These include cellulose derivatives,[90] polyacrylonitrile and plasticised PVC.[91] Attempts have been made to reduce this type of activation by altering the nature of the surfaces. In particular, employing polyethylene oxide grafted polyurethanes has been shown to reduce complement activation.[92] Such surfaces have also been found to be less adherent for platelets.[93] Similarly sulfonate groups have been found to reduce the complement activation induced by hydroxymethyl groups on polystyrene surfaces.[94]

Modification of Blood–Biomaterial Interactions by Heparin

Heparin is a naturally occurring anticoagulant that is produced by mast cells, and it can be used in clinical procedures to modify the response of artificial materials in blood-contacting applications. Its anticoagulant properties arise from its role in activating antithrombin III. However the effect of heparin *in vitro* is not fully understood. It seems to react with complement proteins to influence the activities of proteins and co-factors of the protein system either promoting or inhibiting such activities. Heparin certainly has an effect on complement activation, and one moreover that has been shown to vary with dose.[95] Within the alternative pathway, heparin is able to bind Factor H, a mediating protein of the complement system. Depending on concentration, whether surface-bound or free, and whether the surface itself is activating or non-activating, heparin may either augment or diminish the degree of complement activation. For example, surface-bound heparin can sequester Factor H, which activates complement in the liquid phase but will inhibit the cascade processes at the surface.[96] Heparin potentiates the action of antithrombin III on thrombin, and it is possible that antithrombin III may play a role in the inactivation of convertases,[97] though it also possible that heparin may act directly on the complement, without doing so by way of the potentiation of antithrombin III.[98] Heparin may also interact with C3b and the convertase, which would alter the generation of C3a. Heparin may bind to C3b and block sites for binding of Factor B, required to form the C3a convertase. In the liquid phase, heparin has been reported to influence the interaction of Factors B and D with C3b when in the purified form. Thus, it is apparent that the interaction of heparin with the complement system is far from straightforward.

The interaction of blood with artificial materials in the presence of heparin is modified in a variety of ways. The activation of blood coagulation by artificial materials is so-called contact-phase activation. It seems likely that adsorption of Factor XII is a critical step in the initiation of this process, and when such adsorption occurs on negatively charged surfaces, such as glass or kaolin, it makes Factor XII susceptible to proteolysis and autoactivation.[99] Other proteins are also involved in surface adsorption and coagulation *via* the intrinsic pathway, including prekallikrein and high molecular weight kininogen. Heparin has been reported to activate Factor XII,[100] a feature that seems related to heparin's anionic nature which provides a negatively charged site to initiate contact phase activation. Whether this mechanism occurs has been questioned, and it has certainly been found that, in plasma, heparin does not support contact-phase activation.[101] Heparin has other effects: it can inactivate kallikrein, which is involved in activation of Factor XII,[102] and it may potentiate the activity of C1-Inh, the main inactivator of Factor XIIa by C1-Inh. It is thus apparent that differences in Factor XII activity at a surface are complicated, and are influenced by the chemistry of the material surface, as well as by the interaction between the surface, heparin and other co-factors adsorbed along with Factor XII. Alternatively, these changes can be viewed as the material influencing the interaction of heparin with blood, and the extent to which heparin is able to exhibit its anticoagulant properties.

9 Interactions with Proteins

The interaction of proteins with surfaces is important in determining the biocompatibility of implant materials, because the initial response of the body to the presence of a foreign material almost always involves exposure to blood or extracellular fluid, both of which are rich in proteins. The nature of the interaction of proteins with the surface in turn determines how appropriate cells from the adjacent tissues will respond and this can vary in major ways with only subtle differences in the artificial surface. For example, in one study, two different polystyrene surfaces were used and the protein fibronectin was adsorbed onto them, after which they were exposed to cultured myoblast cells.[103] Although similar amounts of the protein were deposited in each case, the behaviour of the cells varied greatly. In one case, they proliferated, while in the other they differentiated. These differences were attributed to subtle differences in the surfaces of the polystyrenes that led to small but significant differences in the nature of the resulting fibronectin coating, which in turn affected the processes induced in the cells.

Conducting experiments on the subject of protein adsorption poses considerable practical difficulties. Techniques such as X-ray crystallography cannot be used for proteins on surfaces, though of course this technique has been spectacularly successful in determining structures of individual protein crystals. Immunological methods, too, have problems, because the necessary interactions are specific to the particular conformation of the protein. Hence, the lack of a response may indicate one of three possible effects: desorption, denaturation or covering of the protein of interest by a different protein.

Despite these difficulties, a number of facts have been established about the interaction of proteins with artificial surfaces, and any model of the overall process should take account of them. They are:

(i)   Proteins are able to adsorb onto almost all synthetic surfaces. Such adsorption is typically very rapid, and related to the concentration of the protein in the blood plasma.

(ii) Following adsorption, proteins may undergo denaturation, although the degree and rate of this denaturation varies with factors such as pH, ionic strength and concentration of the adjacent solution. Lower concentrations are correlated with greater extents of denaturation, an observation that has been attributed to the longer time available for the conformational changes associated with denaturation to occur before the surrounding sites on the surface become occupied. This assumes that the presence of neighbouring proteins limits the freedom for rearrangement of the original protein molecule, but is consistent with a variety of experimental observations.[104]

(iii) Even under ideal conditions, the resulting layer of protein is heterogeneous, in terms of distribution, orientation and structure, and there are small zones of the surface, too small to accommodate a protein molecule, that remain uncovered.

The simplest approach to modelling the interaction of proteins with surfaces is to consider that single molecules interact with the surface in a manner akin to placing hard billiard balls on a sticky surface.[105] This situation can be described by the equation:

$$\frac{dM}{dt} = c_1 \varphi k_a - M k_d$$

where M is the mass adsorbed per unit area, c_1 is the concentration near the surface, which depends partly on the bulk concentration and partly on hydrodynamic considerations, and k_a and k_d are rate constants for adsorption and desorption respectively. The term φ is the fraction of the surface available for adsorption. Because adsorption takes place at random positions, not at definite sites, φ is not equal to $1 - \theta$, where θ is the fraction of the surface occupied by adsorbed proteins, as would be suggested by the Langmuir model of adsorption. In other words, if every *available* part of the surface were occupied, the surface would not be completely covered by protein molecules because of the existence of exposed areas that are too small for a protein molecule to fit into.

According to the equation, the value of M should rise towards a plateau value as the fresh surface is exposed to a protein solution, and should decay back to zero when the coated surface is exposed to pure solvent. In fact, this is not observed experimentally, but instead what is actually observed is that adsorption is not completely reversible. This suggests that eventually every available site will be occupied by irreversibly bound protein molecules, which in turn suggests that the maximum amount adsorbed should be independent of the initial bulk concentration. Surprisingly, this is rarely observed with real systems. Rather, the higher the bulk concentration in solution, the greater the amount of protein adsorbed,[106] even where the adsorption can be shown to be irreversible.

Three different interpretations have been proposed to account for these observations:

(a) Protein adsorption may occur in at least two distinct orientations. This is supported by the fact that most globular proteins are ellipsoidal, and could in principle adsorb with either the long or the short axis at right angles to the surface (see Figure 6.3). With high bulk concentrations, more molecules land on the surface close to each other, which favours the adsorption mode with the long axis perpendicular to the surface. It also prevents reorientation or relaxation processes following adsorption that might otherwise occur and block neighbouring adsorption sites.

(b) Proteins may form a two-dimensional crystal structure at the surface. Because of the more ordered, close packed nature of the crystal structure, this

Solution

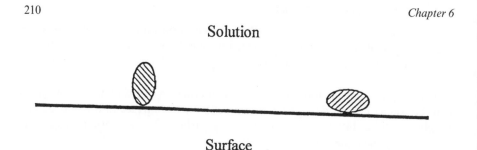

Surface

Figure 6.3 *Possible orientations for adsorption of globular proteins*

results in a greater surface concentration than the more random and relatively disordered array that becomes adsorbed from solutions of lower bulk concentration. This particular behaviour was inferred from the mechanism of adsorption of bovine serum albumin and γ-globulin on polystyrene.[107]

(c) Proteins may denature, and give similar results to the situation described in (a), *i.e.* the protein can relax more readily into a denatured conformation that occupies more area when it is deposited from low concentration solutions than when it is deposited from high concentration solutions. This is because such relaxations are inhibited if neighbouring sites are occupied too rapidly by other protein molecules coming out of solution.

The Vroman Effect

So far we have considered the case of a single protein only adsorbing onto artificial surfaces. In the body, however, this is not the situation: proteins of a great variety of types, functions and concentrations are present, and in order to understand the processes which take place at the surface of an implanted biomaterial we need to consider what happens when a variety of proteins are present. In particular, we need to be aware of the interactions between the proteins themselves, and how these can lead to exchange at the surface and competitive adsorption.

The classic work on this subject is that of Vroman and his co-workers, and the interaction of synthetic surfaces with proteins is sometimes called the *Vroman effect*, though the term has been applied to subtly different phenomena associated with multiple adsorption of protein molecules. The original observation, published by Vroman and Adams in 1969, concerned a glass surface, and was that exposure of this surface to plasma for as little as two seconds led to the deposition of fibrinogen.[108] This fibrinogen was detected by a polyclonal fibrinogen antibody, and was shown to be purely physical and not to involve covalent bond formation.[109] However, after 10 minutes exposure to plasma, this fibrinogen layer was no longer detectable immunologically. Although this result may have been caused by denaturation of the adsorbed protein, it was actually shown to be caused by displacement of the fibrinogen by other plasma proteins.[110]

The displacement of fibrinogen is caused by high molecular weight kininogen (HMWK) in the plasma.[111,112] This is a single chain glycoprotein present in plasma at a concentration of 30–90 μm ml^{-1} and with a molecular weight of approximately 120 000.[113] It occurs as a complex with prekallikrein, and the presence of an artificial surface triggers an elaborate sequence of biochemical events that lead to the cleavage of the complex and a conformational change in the HMWK that favours its deposition onto the surface. The importance of HMWK in the displacement of fibrinogen has been demonstrated in experiments with plasma that is deficient in HMWK. Such plasma shows only the rapid deposition of fibrinogen, but no subsequent loss. Adding HMWK leads rapidly to the loss of fibrinogen at the surface. Simple dilution can also lead to such a low concentration of HMWK that it no longer replaces fibrinogen at the surface.[111,112]

Both the deposition and subsequent loss of fibrinogen and the plasma dilution result have been termed the Vroman effect.[114] The term has also been applied to a third related phenomenon, namely the sequential displacement of any proteins adsorbed from blood or plasma onto an artificial material surface.[115] Initial studies were on glass, but it has since been shown that similar effects occur with metals, *e.g.* germanium, platinum and tantalum, and also silicon[116] and polymers, such as polyethylene[117] or polystyrene.[118] Application of the term *Vroman effect* in this way is its most general use, and as such it is related not only to the nature of the surface concerned but also to the concentration of proteins in the plasma. Initial deposition is of abundant and rapidly adsorbing proteins, of which fibrinogen is the prime but not only example.[119] Displacement occurs due to interaction with more dilute, slower adsorbing proteins. HMWK is particularly effective, but other proteins also behave in this way. For example, Factor-XII is also effective.[118]

Studies have shown that such displacement is complicated, and does not necessarily involve specific displacement in a well-defined sequence.[120] This is because the whole process of surface interaction of proteins involves so many factors. These include the chemistry of the artificial surface, pH and plasma concentration, as well as the concentration of individual proteins within the plasma.[121] Highly dilute plasma does not show the Vroman effect, possibly because the concentration of individual potentially displacing proteins falls below a critical level.[122] The hydophobicity of the surface is also important, and both increases the affinity of the surface for proteins and reduces the rate and extent of displacement.[123]

Flow of plasma across the surface alters aspects of the protein adsorption process, at least quantitatively, but does not change the fact that there is an essentially sequential process.[124,125] This is because of mass transfer effects and hydrodynamic and concentration boundary layers that are set up in a dynamic system. Hence, it is known that these effects must occur *in vivo* as biomaterials are used in service. The details of the process, including the type and conformational state of the protein, are critical factors in determining biocompatibility.[126,127] The rapidity of the initial deposition ensures that the body essentially 'sees' only adsorbed protein within seconds of the implant coming into contact with blood or plasma. It follows, therefore, that it is this deposited protein layer that determines how the body responds and what further biochemical events follow from the formation of this bound protein layer.

10 Cell Adhesion

Most mammalian cells will adhere readily to a surface. It is essential for them to attach to something in order to be able to perform their normal functions, including proliferation and differentiation. Under normal circumstances, the substrate to which they attach is the extracellular matrix, ECM, a biological support structure that consists of insoluble proteins and glycoaminoglycans.[128] In addition to acting as a mechanical support for cells, ECM contains numerous peptide and carbohydrate ligands that are recognised by cellular receptors, and interactions between these ligands and the cells are vital to maintain proper cell function and to allow cells to respond appropriately to their environments. The main function of the ECM, though, is to provide a support for the adhesion of cells.[129] Unless they can adhere to a substrate, most cells undergo a process of apoptosis, or programmed cell death, and simply cease to function. In addition, the loss of signal transduction pathways associated with adhesion leads to growth and spreading of tumours.

As has already been mentioned, proteins readily deposit onto surfaces, and it is to the resulting protein coating that cells actually adhere. There are interactions between the protein layer and ligands from the ECM that mediate behaviour of cells on surfaces, and the nature and extent of the protein coating can have profound effects on what the cells do, including whether they proliferate or differentiate.

For most cells, adhesion is regulated by a class of cell receptors called integrins.[130] These substances comprise about twenty heterodimeric proteins, and they are involved in many aspects of cell behaviour, not only adhesion, but also migration, proliferation and differentiation. Studies in this area are complicated because most cells have a variety of integrins on their surfaces, and also because the ECM has many ligands that will interact with these integrins. The range of possible interactions is therefore large. There have been useful insights obtained from experiments on model substrates containing only a single ligand, for example, the tripeptide Arg-Gly-Asp. This ligand is found in fibronectin and is involved in interactions with several integrin receptors. When materials are derivatised with this peptide, an enhancement has been found in the growth and attachment of cells.[131] However, this peptide alone is not sufficient for complete spreading of cells. Additional experimental work using kidney cells from immature hamsters and substrates coated with fragments and mutants of fibronectin have shown that a second peptide, Pro-His-Ser-Arg-Asp, must be present adjacent to the Arg-Gly-Asp peptide if cells are to attach and spread completely on the surface.[132] Studies of this kind have shown that both attachment and spreading of cells are improved in the presence of appropriate ligands on the surface, and have begun to shed light on the way in which such attachment of cells varies with ligand type.[133]

Adhesion of Osteoblasts

The phenomenon of osseointegration exhibited by certain metallic alloys when used as implants in orthopaedic surgery and in dentistry relies on close growth of bone towards the implant with the formation of fibrous capsule. This occurs in part because of the capability of osteoblast cells to adhere to these materials, typically titanium alloys, and to spread, grow and differentiate without adverse effects.

The interaction between osteoblasts and materials depends on the detail of the surface composition, surface energy and topography. These features determine not only how the biological molecules that promote adhesion will adsorb, but also how they will orientate.[134] This in turn will affect how the cells spread, and studies of cell migration on biomaterials with gross surface features, such as grooves, have demonstrated that migration tends to be more rapid on those surface with deeper grooves.[135]

Surface chemistry has been shown to have effects in a variety of studies and with a range of biomaterials. For example, sterilisation of commercially pure titanium has been shown to alter the detailed chemistry of the surface[136] and these changes in turn modify subsequent cell adhesion.[137,138] In particular the surface oxide layer can be altered adversely by inappropriate sterilisation techniques, and this in turn can alter the reaction of cells to the resulting surface.[139]

As we have seen, cells are not presented with a pure material surface when a device is implanted within the body. Any such surface rapidly becomes covered with a layer of protein molecules, and, though the parent surface strongly influences the nature of such adsorption,[140] the fact remains that it is the protein layer which confronts the cells, not the synthetic material surface.[141] Cell adhesion is generally enhanced by hydrophilic surfaces and the adsorption of proteins such as vitronectin[142,143] and plasma fibronectin causes such surfaces to develop,[144] thereby promoting cell adhesion. The presence of vitronectin in particular is critical for the spatial distribution, attachment and spreading of osteblasts and other bone cells on positively charged regions of surfaces.

The adhesion of osteoblasts is related to the expression of various adhesion proteins. In cultures of osteoblasts, the proteins actin and vinculin have been shown to be present using immuno-histochemical techniques. Cell adhesion has been found to follow the order Ti6Al4V > polystyrene > CoCrMo, and this was linked to the observation that Ti6Al4V showed an enhanced rate of actin and vinculin cytoskeletal reorganisation. In cell cultures of osteoblasts in contact with Ti6Al4V, vinculin was shown to form short, thin and dense patches,[145] and such surfaces appear optimal for osteoblast attachment.

Once the cells have become attached, they synthesise extracellular matrix proteins. For example, neonatal rat calvarial osteoblasts have been shown to express osteonectin and osteopontin when cultured on Ti6Al4V, hydroxyapatite or tissue culture polystyrene.[146] The details of bone protein expression are affected by the substrate to which the cells are attached. For example, rat bone marrow stromal cells expressed more osteopontin and bone sialoprotein on hydroxyapatite than on cpTi or a glass-ceramic,[147] and human osteoblasts synthesised greater quantities of proteins on CoCrMo alloy than on glass discs.[148]

Surface roughness also influences the synthesis of proteins by adsorbed cells, though in variable directions. Using human osteoblast-like cells (MG63) on titanium discs, it has been shown that increasing roughness reduced collagen synthesis without affecting the synthesis of non-collagenous proteins.[149] On the other hand, production of PGE-2 and TGFβ_1 was higher on rough surfaces than on smooth ones.[150] Other aspects of cell biochemistry were influenced by surface roughness as well: as roughness increased so did osteocalcin synthesis and alkaline phosphatase activity.[151]

As the process of cell adhesion is becoming better understood, so there is scope for improving biomaterials. The possibility exists of adsorbing the relevant proteins onto the surface prior to implantation, with the aim of enhancing cellular attachment and possibly preventing the formation of fibrous capsule. Preliminary studies using synthetic peptides have shown the potential of this approach,[152] though it is very early days and no surgical devices are yet available with this kind of surface pretreatment.

11 Interactions with Hard Tissues

The hard tissues of the body consist of the teeth and bones, but the detailed structure of the teeth components, enamel and dentine, and of the various types of bone within the body varies considerably. Consequently, so does their interaction with artificial materials.

Teeth are composed of two parts, the outer enamel and the inner dentine.[153] Enamel is the hardest substance in the body, and consists mainly of mineral, hydroxyapatite, with minor amounts of organic matter. It is acellular, and hence not considered to be tissue. Dentine, by contrast, is softer and consists of a greater proportion of organic matter, mainly collagen and other protein. Structurally, it has a series of tubules running through it to the pulp, the chamber of the tooth that contains the nerve and blood supply. It is cellular, and hence considered to be properly a tissue of the body.

The interaction of teeth with synthetic materials tends not to involve much if any biological activity. However, certain materials have been shown to promote remineralisation in the dentine, most notably the glass-ionomer dental cement. Studies have shown the development of a 'zone of interaction' when this material remains in the tooth for up to five years. This zone, which is distinct when viewed with the electron microscope, contains elements from both the restorative material and the tooth.[154] On the other hand, the more widely used dental restorative materials, such as amalgam, or the so-called composite resins, show no such interaction, and either do not bond to the tooth surface or do so with the aid of substances that act entirely by physico-chemical mechanisms.[155]

Bone represents a much greater volume of the hard tissue of the body than dentine or enamel. It is a living material made up of cells, well supplied with blood, and has three major components, namely collagen, a tough flexible proteinaceous substance; hydroxycarbonate apatite, the bone mineral, which acts as the reinforcing phase; and bone matrix or ground substance, which performs various support functions. These components are assembled so that the resulting three-dimensional system exhibits its maximum mechanical properties along the lines of applied stress.[156]

Bone that needs replacing with an artificial material is usually structurally weak, because of either disease or ageing. Disease and age can also adversely affect the rate of repair of bone.[157] Unfortunately, the quality of bone at the interface with an implant can deteriorate further, either due to the presence of the implant or because of the method of fixation. For example, acrylic bone cements based on poly(methyl methacrylate) undergo a considerable exotherm as they harden, and the rise in temperature can kill bone cells to a depth of almost a millimetre.

The presence of an implant causes changes in the loading pattern in the region of the device, a phenomenon known as *stress shielding*. Bone, however, must be loaded in order to remain in place and, under conditions of clinical service, the changes in load distribution arising from the presence of the implant cause the bone to change shape, so-called remodelling. The osteoblasts themselves are sensitive to mechanical stimuli, and this is manifested in changes in proliferation and possibly phenotypic expression.[158] Mechanical loading also influences adhesion of cells to artificial surfaces. All of these phenomena are associated with differences in protein production by cells which are subject to strains. Under conditions of mechanical loading, osteopontin synthesis generally increases,[159] but both osteocalcin synthesis[160] and collagen expression[161] vary according to the cell type and line.

The cytoskeleton is also affected by mechanical stimulation. Periodic compression of cells has been found to increase actin expression, and mechanical strain has been shown to increase the formation and thickening of actin stress fibres.[162] Mechanical stimulation has also been found to lead to marked increases in vinculin production, especially towards the edges of the cells,[163] though this is not thought to be important in mechanical transduction because there is evidence that depletion of vinculin does not interfere with the response of cells to mechanical loading.[164] Mechanical strains have also been found to enhance the synthesis of prostaglandin E2,[165] and this in turn stimulates the production of cAMP, a protein involved in signal transduction.

Mechanical loads are detected in cells mainly through a matrix–integrin mechanosensory protein complex, and this is linked to a kinase cascade system. The mechanosensory complex includes talin, vinculin, tensin, paxillin, Src and focal adhesion kinase, FAK. Following stimulation, and in timescales in the region of milliseconds to seconds, a complex signalling response is initiated that involves various ions (Ca^{2+}, Na^+, K^+, H^+) and proteins (IP3, cAMP, kinases *etc.*). Some time (minutes to hours) after this initial stimulation, signalling with kinases, transcription and transduction occurs, proteins of the cytoskeleton undergo polymerisation, and focal adhesion rearrangements take place. Then, over a period of days, cells migrate, express and degrade ECM, divide or die, thus establishing a new state of equilibrium.[166] This new equilibrium results in bone being maintained in a new shape, built up in regions of increased stress and depleted (or even removed altogether) from regions of reduced stress.

As the new configuration of bone develops, so the interface with the implant deteriorates. This can result in various undesirable outcomes: loosening, fracture of the bone or failure at the interface. The presence of wear debris from prosthetic hip or knee joints can accelerate the weakening of stress-shielded bone, because wear particles stimulate the cells into increased activity. This may provoke attack on the aligning bone, resulting in its destruction. Thus stress shielding in association with the presence of wear debris can be highly damaging, and may lead to failure of the implant.[167]

Among the biochemical species involved in bone resorption are cytokines. This is the term applied to soluble immunological factors discovered in the 1960s, that are produced by various cells, but mainly by T-cells and macrophages. Cytokines are generally distinct from other growth factors which have non-immune cells as their target for activity.

Chemically, cytokines are relatively low molecular weight (6–66 000) glycoproteins. They are very potent, and are active in the concentration range 1×10^{-9} to 1×10^{-12} g cm^{-3}. The expression of cytokines by cells is regulated by a delicate and complex array of receptors and signalling molecules, and production is restricted to those cells that have been activated by some kind of biochemical signal. Cytokines act to modulate the activities of a variety of target cells, and a particular cytokine may be responsible for stimulation or inhibition, depending on concentration. Effects may vary depending on the state of the target cells, and also by the order in which various cytokines are able to reach the target cells.

It is a group of pro-inflammatory cytokines, including IL-1β,[168] which is involved in bone resorption. IL-1β is found in activated phagocytes and is also produced by human adult bone *in vitro*.[169] This substance increases the synthesis of prostaglandins in bone; it may also be involved in osteoclast differentiation.[170] The production of tumour necrosis factor α, TNFα, is stimulated by IE1β and this substance can also enhance the formation of osteoblast-like cells in bone marrow culture.[171]

Cytokines of the Interleukin 6 (IL-6) type are known to affect skeletal homeostasis, which they do by influencing the development and function of bone cells. IL-6 may regulate its own production as well as regulating the components of its receptor, which may in turn influence the rate of bone remodelling.[172] Another substance involved is granulocyte monocyte-colony stimulating factor, GM-CSF, which is produced by a variety of cells, including fibroblasts and osteoblasts. This substance acts on precursors to osteoblasts, thereby stimulating bone resorption. This is probably due to the activation of macrophages by GM-CSF, which causes them to release the prostaglandins and other cytokines responsible for stimulating bone resorption.[173]

A satisfactory implant would generally be one that became fully integrated into the bone, with bone growing right up against the artificial material, so-called osseointegration. By comparison, a less satisfactory implant is one that promotes formation of fibrous capsule. These two possible responses are now considered in more detail.

Osseointegration

Osseointegration is defined as a direct bone anchorage to an implant that has the ability to transmit loading forces directly to the bone.[174,175] This means that the implant becomes fully integrated into functioning bone, with no fibrous capsule interface.

The discovery of osseointegration was made by Branemark at the University of Göteborg, Sweden, as a result of studies of bone repair mechanisms. In his experiments, he inserted a titanium chamber surgically into the tibia of rabbit. Unexpectedly, bone developed right up against this chamber and bonded to it strongly.[176] Following this, he went on to develop implants for tooth roots based on titanium,[177,178] as well as carrying out fundamental studies on the basic biology of osseointegration.[179,180]

As a result of these experimental studies, and also following from many years of clinical observation, the following key factors have been identified for successful osseointegration:

(1) *Choice of material.* Essentially, osseointegration requires the use of titanium-based implants, such as cpTi. In air, this develops an oxide layer 500–1000 nm thick,[181,182] but once the implant has healed into the bone, this layer becomes surrounded by further layers, one of glycoprotein, the other of calcified tissue.[183]

(2) *Design of device.* The device must be fabricated with a screw thread. This serves two purposes: firstly, it increases the overall surface area against which bone can grow, and secondly it balances the distribution of forces into the surrounding bone.[184,185] The thread enables the implant to achieve secure anchorage immediately, and this prevents any micromovement, which would otherwise cause fibrous capsule to be formed.

(3) *Prevention of excessive heat generation during bone drilling.* It is essential that bone does not become heated above 43 °C, since this is the temperature at which alkaline phosphatase begins to break down.[186] Generally, surgeons using this technique try to keep the temperature below 40 °C, which is done using gentle pressure with the surgical drill, and copious amounts of sterile saline irrigation.

(4) *Avoidance of premature loading.* The normal clinical approach with osseointegrated implants is to leave them undisturbed for six months in the maxilla and for 3–4 months in the mandible.[187] This enables the implant to become fully integrated into the bone before loads are applied, and takes

account of the timescale of the development of mature healthy bone. Bone healing begins almost immediately the implant has been placed, certainly within the first week, and reaches a peak at 3–4 weeks. The initially deposited tissue gradually calcifies after 6–8 weeks, and this is followed by full remodelling to healthy cortical bone.

Fibrous Capsule Formation

Fibrous capsule can be formed in a variety of circumstances, for example around tumours in soft tissue or around air bubbles injected into muscle.[188] Of particular importance is that it can be formed in bone, for example around surfaces of fractured bone that are subject to motion. The formation of such fibrous capsule may be induced by macrophage activity, these macrophages having been activated by the foreign body reaction. The fibrous capsule may mediate bone resorption *via* production of prostaglandin E_2,[189] though PGE_2 has also been assumed to be a promoter of bone formation in other studies. Any such bone resorption would lead to aseptic loosening of the implant. Within such fibrous capsule, and in bone around artificial materials, certain cells produce synovial fluid. These cells are known as synoviocytes, and seem to develop from stem cells within the fibrous capsule, rather than migrating into the capsule from outside.

Fibrous capsule has been shown to contain a variety of cells, including macrophages, lymphocytes and mast cells that are associated with an inflammatory reaction.[190] These cells produce a variety of chemical mediators, cytokines and growth factors which modulate the inflammatory process.

Recently the possibility that apoptosis may occur in hard tissues surrounding orthopaedic implants has been studied.[191] Apoptosis is controlled by intracellular proteins, being promoted by bax but inhibited by bcl-2.[192,193] By employing immuno-histochemical techniques to monitor these proteins and detecting DNA fragments to confirm cell death, Zhang and Revell[191] were able to demonstrate that apoptosis does, indeed, occur in the fibrous capsule that forms due to aseptic lossening of prosthetic hip and knee implants.

12 Safety Testing of Biomaterials

Clearly the issue of safety underpins all of the studies of the biological interactions of artificial materials. However, there are also International Standards for the testing of medical devices[194] and their role is more than to confirm biological safety. They have a legal status and, for example, products can be sold within Europe only of they bear the CE mark, which may be obtained mainly be having passed the relevant, recognised Standard tests.

The international Standard for the biological evaluation of medical devices is ISO 10993, which is more or less identical with EN 30993, and for which the details are listed in Table 6.8.

Within Europe, there are now European Standards (ENs) that cover these topics, with the exception of Part 8, and these arose from the harmonised

Table 6.8 *Topics covered by ISO 10993: Biological evaluation of medical devices*

Part	Title
1	Guidance on selection of tests
2	Animal welfare requirements
3	Tests for genotoxicity, carcinogenicity and reproductive toxicology
4	Selection of tests for interaction with blood
5	Tests for cytotoxicity: *in vitro* methods
6	Tests for local effects after implantation
7	Ethylene oxide sterilisation residues
8	Clinical investigation
9	Degradation of materials related to biological testing
10	Tests for irritation and sensitisation
11	Tests for systemic toxicity
12	Sample preparation and reference materials

legislative system that came into force in the early 1990s.[195] To date, two Directives have been issued relating to medical devices, namely the Directive on Active Implantable Medical Devices (90/385/EEC) and the Directive on Medical Devices (93/42/EEC). Once adopted, such Directives are designed to be incorporated into the national law of each member state, bringing about Europe-wide harmonisation of legislation and the removal of technical barriers to trade.

The aim is that eventually these EU Directives will replace existing national systems for the regulation of product safety, though it will remain possible for national Competent Authorities to authorise the marketing of specific products within individual countries which have not complied with the required conformity assessment procedures, on the understanding that their use is in the interest of protection of health.

Directives include so-called essential requirements, and these describe the safety aspects which have to be met before the CE mark can be awarded, and the product sold throughout Europe. Four main aspects for safety of medical devices are considered, namely:

(i)   chemical (mainly toxicological) safety;
(ii) microbiological safety;
(iii) physical safety;
(iv) biological safety.

This final requirement relates to the biocompatibility of the product and is, in fact, a consequence of the first three items. Nonetheless, it is referred to explicitly and assumes that particular attention will be paid to the compatibility of the materials used with the biological tissues, cells and body fluids in which they are to be placed. These requirements also include the stipulations that:

Table 6.9 *Additional tests undergoing development for EN 30093*

Part	Title
13	Identification and quantification of degradation products from polymers
14	Identification and quantification of degradation products from ceramics
15	Identification and quantification of degradation products from coated and uncoated metals and alloys
16	Toxicokinetic study design for degradation products and leachables
17	Material characterisation

(i) particular attention must be paid to the tissues exposed and the duration and frequency exposure;

(ii) the device must be designed and manufactured in such a way as to reduce to a minimum the risks posed by substances leaking from the device.

There are a number of ways in which manufacturers can confirm that these essential requirements are met by their products, and thus obtain authority to display the CE mark on them. Whichever route is chosen, the application of Standards is involved. How stringent the tests are depends on which Directive applies, and on the classification of the particular product. Classification and consequent level of control are related to the degree of risk associated with the particular medical device. These are formally classified as: Class I (low risk), Class IIa and IIb (medium risk) and Class III (high risk). For Classes IIa, IIb and III, manufacturers must have the assessment of conformity procedure carried out by a national competent authority, a so-called Notified Body.

The testing procedures employed essentially relate to those developed under ISO 10993, and listed in Table 6.8. However, for these European Standards in the series designated EN 30093, additional tests have been identified and these are listed in Table 6.9.

The evaluation of medical devices to demonstrate conformity with essential requirements is a complex procedure, since both safety and effectiveness need to be considered, and Standards are central to the process. Such Standards tend to be classified as:

(i) horizontal, *i.e.* those which cover general safety and performance criteria;

(ii) vertical, *i.e.* those which are specific to a product or group of products.

The question of biological testing of medical devices is related to risk analysis,[196] and the current legal framework, particularly in Europe, aims to strike a balance between the desired degree of safety and the introduction of further risks. It is recognised as important for the general improvement in public health that development costs for new materials and devices must be contained within reasonable limits. In ensuring that toxicological risks are acceptable, it is also recognised that absolute safety is impossible and that no material can be guaranteed to be fully biocompatible. The aim instead is to achieve a realistic level of assurance that toxicity of the material has been adequately controlled. Such a balanced approach, placing known risks in a realistic context, should prove to be the basis for a reliable and cost-effective means of minimising the problems of adverse biological effects without stifling innovation and the search for new and improved materials and devices. That is certainly the aim.

13 References

1   D. F. Williams, *Progress in Biomedical Engineering, Definitions in Biomaterials*, Vol. 4, Elsevier, Amsterdam, 1987, p. 954.
2   C. J. Kirkpatrick, F. Bittinger, M. Wagner, H. Köhler, T. G. van Kooten, C. L. Klein and M. Otto, *Proc. Inst. Mech. Eng., Part H*, 1998, **212**, 75.
3   P. Thomsen, C. Larsson, L. E. Ericson, L. Sennerby, J. Lausmaa and B. Kasemo, *J. Mater. Sci.; Mater. Med.*, 1997, **8**, 653.
4   B. D. Ratner, *J. Biomed. Mater. Res.*, 1993, **27**, 837.
5   D. F. Williams, *Concise Encyclopaedia of Medical and Dental Materials*, xvii, Pergamon Press, Oxford, UK, 1990.
6   D. F. Williams, *J. Mater. Sci.*, 1987, **22**, 3421.
7   D. Bakker, C. S. van Blitterswijk, S. C. Hesseling, J. J. Grote and W. T. Dries, *Biomaterials*, 1988, **9**, 14.
8   A. Dekker, C. Panfil, M. Valdor, G. Pennartz, H. Richter, Ch. Mittermayer and C. J. Kirkpatrick, *Cells and Mater.*, 1994, **4**, 101.
9   C. J. Clifford and S. A. Downes, *J. Mater. Sci.; Mater. Med.*, 1996, **7**, 637.
10   H. G. Gratzner, *Science*, 1982, **218**, 474.
11   C. J. Kirkpatrick and A. Dekker, *Adv. Biomater.*, 1992, **10**, 31.
12   J. L. Duval, R. Warocquier-Clerout and M. F. Sigot-Luizard, *Cells Mater.*, 1992, **2**, 179.
13   J. L. Duval, R. Warocquier-Clerout and M. F. Sigot-Luizard, *Cells Mater.*, 1995, **5**, 1.
14   W. N. Aldridge, in D. Anderson and D. M. Conning, *Experimental Toxicology*, 2nd Edition, Royal Society of Chemistry, Cambridge, UK, 1993, Chapter 5.
15   F. O. Brady and R. L. Kafka, *Anal. Biochem.*, 1979, **98**, 89.
16   J. J. Jacobs, A. K. Skipor, J. Black, R. M. Urban and J. O. Galante, *J. Bone Jt. Surg.*, 1991, **73A**, 1475.
17   M.-F. Hammond, *Biocompatibility Assessment of Medical Devices and Materials*, ed. J. H. Braybrook, John Wiley & Sons, Chichester, UK, 1997, Chapter 6, Part 1, pp. 119–124.
18   B. L. Dahl and L. Tronstad, *J. Oral Rehab.*, 1976, **3**, 19.
19   P. H. Krammer, I. Behrmann, P. Daniel, J. Dhein and K. M. Debarin, *Curr. Opin. Immunol.*, 1994, **6**, 279.
20   G. Majno and F. Joris, *Am. J. Pathol.*, 1995, **146**, 3.
21   C. A. Dinarelo, *Blood*, 1996, **87**, 2095
22   S. Nagata, *Cell*, 1997, **88**, 355.
23   G. Kroemer, *Nature Med.*, 1997, **3**, 614.
24   Y. Soini, P. Paakko and V. P. Lehto, *Am. J. Pathol.*, 1998, **153**, 1041.

25　D. C. S. Huang, L. A. O'Reilly, A. Strasser and S. Cory, *Eur. Mol. Biol. Org. J.*, 1997, **16**, 4628.

26　N. Hallab, J. J. Jacobs and J. Black, *Biomaterials*, 2000, **21**, 1301.

27　D. A. Basketter, G. Briatico-Vangosa, W. Kaestner, C. Lally and W. J. Bontinck, *Contact Dermatitis*, 1993, **28**, 15.

28　C. Liden, J. E. Wahlberg and H. I. Maibach, *Skin*, in *Metal Toxicology*, ed. R. A. Goyer, C. D. Klaassen and M. P. Waalkes, Academic Press, New York, 1995, pp. 447–464.

29　P. Haudrechy, J. Foussereau, B. Mantout and B. Baroux, *Contact Dermatitis*, 1994, **31**, 249.

30　J. Yang and K. Merritt, *J. Biomed. Mater. Res.*, 1996, **31**, 71.

31　C. Angle, in *Metal Toxicology*, ed. R. A. Goyer, C. D. Klaassen and M. P. Waalkes Academic Press, New York, 1995, pp. 71–110.

32　P. A. Lalor, P. A. Revell, A. B. Gray, S. Wright, G. T. Railton and M. A. R. Freeman, *J. Bone Jt. Surg.*, 1991, **73B**, 25.

33　A. W. Parker, D. Drez Jr and J. J. Jacobs, *Am. J. Knee Surg.*, 1993, **6**, 129.

34　H. I. Abdallah, R. K. Balsara and A. C. O'Riordan, *Ann. Thoracic Surg.*, 1994, **57**, 1017.

35　R. H. M. Thomas, M. Rademaker, N. J. Goddard and D. D. Munro, *Br. Med. J.*, 1987, **294**, 106.

36　C. Merle, M. Vigan, D. Devred, P. Girardin, B. Adessi and R. Laurent, *Contact Dermatitis*, 1992, **27**, 257.

37　E. Spiechowitz, P. Glantz, T. Axell and W. Chmielewski, *Contact Dermatitis*, 1984, **10**, 206.

38　L. Kanerva, T. Sipilainen-Malm, T. Estlander, A. Zitting, R. Jolanki and K. Tarvainen, *Contact Dermatitis*, 1994, **31**, 249.

39　J. Yang and K. Merritt, *J. Biomed. Mater. Res.*, 1994, **28**, 1249.

40　M. Cramers and U. Lucht, *Acta Orthop. Scand.*, 1977, **48**, 245.

41　J. Black, *Biomaterials*, 1984, **5**, 12.

42　J. Foussereau, *Trans. St John's Hosp. Dermatol. Soc.*, 1966, **52**, 220.

43　M. H. Samitz and A. Klein, *J. Am. Med. Assoc.*, 1973, **223**, 1159.

44　P. M. Gordon, M. I. White and T. R. Scotland, *Contact Dermatitis*, 1994, **30**, 181.

45　K. Merritt and S. Brown, *Int. J. Dermatol.*, 1981, **20**, 89.

46　V. P. Barranco and H. Solloman, *J. Am. Med. Assoc.*, 1972, **220**, 1244.

47　G. Rostoker, J. Robin, O. Biret *et al.*, *J. Bone Jt. Surg.*, 1987, **69A**, 1408.

48　K. Merritt and S. Brown, *Acta Orthop. Scand.*, 1980, **51**, 403.

49　R. S. Cotran, *Robbins' Pathologic Basis of Disease'* 5th Edition, ed. V. Kumar and S. L. Robbins, W. B. Saunders, Philadelphia, PA, 1994.

50　F. J. Schoen, *Trans. Am. Soc. Artif. Organs*, 1987, **33**, 8.

51　W. McDonald, *Cancer*, 1981, **48**, 1009.

52　J. J. Ward, W. K. Dunham, D. D. Thornbury and J. E. Lemons, *Trans. Soc. Biomater.*, 1987, **10**, 106.

53　J. Bago-Granell, M. Aguirre-Canyadell, J. Nardi and Talled, *J. Bone Jt. Surg.*, 1984, **66B**, 38.

54　R. K. N. Ryu, E. G. Bovill Jr, H. B. Skinner and W. R. Murray, *Clin. Orthop. Rel. Res.*, 1987, **216**, 207.

55　A. Martin, T. W. Bauer, M. T. Manley and K. H. Marks, *J. Bone Jt. Surg.*, 1988, **70A**, 1561.

56　J. Lamovec, A. Zidar and M. Cucek-Plenicar, *J. Bone Jt. Surg.*, 1988, **70A**, 1558.

57　P. C. Weber, *J. Bone Jt. Surg.*, 1986, **68B**, 824.

58   D. S. Weinberg and B. S. Maini, *Cancer*, 1980, **46**, 398.
59   J. W. Fehrenbacker, W. Bowers, R. Strate and J. Pittman, *Ann. Thorac. Surg.*, 1981, **32**, 297.
60   K. G. Brand, L. C. Buoen, K. H. Johnson and I. Brand, *Cancer Res.*, 1975, **35**, 279.
61   B. S. Oppenheimer, E. T. Oppenheimer and A. P. Stout, *Proc. Soc. Exp. Biol. Med.*, 1948, **67**, 33.
62   F. W. Sunderman, *Fund. Appl. Toxicol.*, 1989, **13**, 205.
63   J. J. Jacobs, D. H. Rosenbaum, R. M. Hay, S. Gitelis and J. Black, *J. Bone Jt. Surg.*, 1992, **74**, 740.
64   A. J. Aboulafia, K. Littleton, B. Smookler and M. M. Malawer, *Orthop. Rev.*, 1994, **23**, 427.
65   M. Spector, C. Cease and X. Tong-Li, *Crit. Rev. Biocompatibility*, 1989, **5**, 269.
66   G. Weissman, J. E. Smolen and H. M. Korchak, *New Engl. J. Med.*, 1980, **303**, 27.
67   P. M. Henson, *J. Immunol.*, 1971, **107**, 1547.
68   J. M. Anderson and K. M. Miller, *Biomaterials*, 1984, **5**, 5.
69   D. K. Han, S. Y. Jeong, Y. H. Kim and B. G. Min, *J. Biomed. Sci., Polym. Edn.*, 1992, **3**, 229.
70   V. Sa Da Costa, D. Brier-Russell, E. W. Saltzmann and E. W. Merrill, *J. Coll. Interface Sci.*, 1981, **80**, 445.
71   R. G. Mason, R. W. Shermer and N. F. Rodman, *Am. J. Pathol.*, 1972, **69**, 271.
72   V. T. Turitto and H. J. Weiss, *Science*, 1980, **207**, 541.
73   B. Ghebrehiwet, B. P. Randazzo, J. T. Dunn, M. Silverberg and A. P. Kaplan, *J. Clin. Inv.*, 1983, **71**, 1450.
74   R. J. Johnson, *Biomater. Sci.*, 1996, 173.
75   M. K. Pangburn and H. J. Muller-Eberhard, *J. Exp. Med.*, 1980, **152**, 1102.
76   M. S. Payne and T. A. Horbett, *J. Biomed. Mater. Res.*, 1987, **21**, 843.
77   N. A. Hoenich, C. Woffindin, S. Stamp, S. J. Roberts and J. Turnbull, *Biomaterials*, 1997, **18**, 1299.
78   P. R. Craddock, J. Fehr, A. P. Dalmaso, K. L. Brigham and H. S. Jacob, *J. Clin. Inv.*, 1977, **59**, 879.
79   D. E. Chenowith, *Artif. Organs*, 1984, **8**, 281.
80   R. Macnair, M. J. Underwood and G. D. Angelini, *Proc. Inst. Mech. Eng., Part H*, 1998, **212**, 465.
81   B. Kasemo and J. Lausmaa, *Environ. Health Perspect.*, 1994, **102**, 41.
82   B. D. Ratner, A. B. Johnson and T. J. Lenk, *J. Biomed. Mater. Res., Appl. Biomater.*, 1987, **21**, 59.
83   T. Beugeling, *J. Polym. Sci., Polym. Symp.*, 1979, **66**, 4129.
84   B. D. Ratner, *J. Biomed. Mater. Res.*, 1993, **27**, 283.
85   G. H. M. Engbers, L. Dost, W. E. Hennink, P. A. M. M. Aaarts, J. J. Sixma and J. Feijen, *J. Biomed. Mater. Res.*, 1987, **21**, 613.
86   K. Park, F. W. Mao and H. Park, *J. Biomed. Mater. Res., Appl. Biomater.*, 1989, **23(A2)**, 211.
87   C. L. Haycox and B. D. Ratner, *J. Biomed. Mater. Res.*, 1993, **27**, 1181.
88   G. M. Bernacca, M. J. Gulbranson, R. Wilkinson and D. J. Wheatley, *Biomaterials*, 1998, **19**, 1151.
89   K. Hayashi, H. Fukumara and N. Yamamoto, *J. Biomed. Mater. Res.*, 1990, **24**, 1385.
90   S. Sundaram, L. Irvine, J. M. Courtney and G. D. O. Lowe, *Int. J. Artif. Organs*, 1991, **14**, 729.
91   N. M. K. Lamba, PhD thesis, University of Strathclyde, Glasgow, UK, 1994.

92 D. K. Han, K. D. Park, K.-D. Ahn, S. Y. Jeong and Y. H. Kim, *J. Biomed. Mater. Res., Appl. Biomater.*, 1989, **23(A1)**, 87.

93 D. K. Han, S. Y. Jeong and Y. H. Kim, *J. Biomed. Mater. Res., Appl. Biomater.*, 1989, **23(A2)**, 211.

94 B. Montdargent, F. Maillet, M. P. Carreno, M. Jozefowicz, M. Kazatchkine and D. Labarre, *J. Biomed. Mater. Res.*, 1990, **24**, 1385.

95 T. Takaoka, P. S. Malchesky and Y. Nose, *Prog. Artif. Organs*, 1986, 553.

96 M. D. Kazatchkine, D. T. Featon, J. E. Silbert and K. F. Austin, *J. Exp. Med.*, 1979, **150**, 1202.

97 J. M. Weller and R. J. Linhargt, *J. Immunol.*, 1991, **146**, 3889.

98 F. Maillet, M. Petiton, J. Choay and M. D. Katzatchkine, *Mol. Immunol.*, 1988, **25**, 917.

99 J. H. Griffen, *Proc. Natl. Acad. Sci. USA*, 1978, **75**, 1998.

100 Y. Hojima, C. G. Cochrane, R. C. Wiggins, K. F. Austin and R. L. Stevens, *Blood*, 1984, **6**, 1453.

101 P. A. Pixley, A. Cassello, R. A. Cadena, N. Kaufmann and R. W. Colman, *Thromb. Haemostat.*, 1991, **66**, 540.

102 G. Fuhrer, M. J. Gallimore, W. Heller and H.-E. Hoffmeister, *Blut*, 1990, **61**, 258.

103 A. J. Garcia and D. Boettiger, *Mol. Biol. Chem.*, 1999, **10**, 785.

104 R. R. Siegel, P. Harder, R. Dahint, M. Grunze, F. Josse, M. Mrksich and G. M. Whitesides, *Anal. Chem.*, 1997, **69**, 3321.

105 J. J. Ramsden, *Chem. Soc. Rev.*, 1995, **24**, 73.

106 S. P. Palecek, J. C. Loftus, M. H. Ginsberg, D. A. Lauffenburger and A. F. Horwitz, *Nature*, 1997, **385**, 537.

107 B. D. Fair and A. M. Jamieson, *J. Coll. Interface Sci.*, 1980, **77**, 525.

108 L. Vroman and A. L. Adams, *J. Biomed. Mater. Res.*, 1969, **3**, 43.

109 L. Tang, C. Tsai, W. W. Gerberich and D. R. Kania, *Biomaterials*, 1995, **16**, 483.

110 J. L. Brash and P. ten Hove, *Thromb. Haemostas.*, 1984, **51**, 326.

111 H. Elwing, A. Askendal and I. Lundstrom, *J. Biomed. Mater. Res.*, 1987, **21**, 1023.

112 A. Poot, T. Beugeling, W. G. Van Aken and A. Bantjes, *J. Biomed. Mater. Res.*, 1990, **24**, 1021.

113 R. M. Cornelius and J. L. Brash, *Biomaterials*, 1999, **20**, 341.

114 T. A. Horbett, *Thromb. Haemostas.*, 1984, **51**, 174.

115 J. L. Brash, in *Modern Aspects of Protein Adsorption on Biomaterials*, ed. J. Y. F. Missirlis and W. Lemm , Kluwer, Dordrecht, The Netherlands, 1991, p. 21.

116 R. E. Baier and R. C. Dutton, *J. Biomed. Mater. Res.*, 1969, **3**, 191.

117 P. Turnbull, T. Beugeling and A. A. Poot, *Biomaterials*, 1996, **17**, 1279.

118 R. J. Green, M. C. Davies, C. J. Roberts and S. J. B. T. Tendler, *Biomaterials*, 1999, **20**, 385.

119 L. Vroman and A. L. Adams, *J. Colloid Interface Sci.*, 1986, **111**, 391.

120 W. Breemhaar, E. Brinkman, D. J. Ellens, T. Beugeling and A. Banjes, *Biomaterials*, 1984, **5**, 269.

121 T. A. Horbett, *ACS Symp. Series*, 1995, **602**, 1.

122 J. D. Andrade (ed.), in *Protein Adsorption, Surface and Interfacial Aspects of Biomedical Polymers*, Vol. 2, Plenum Press, New York, 1985, pp. 1–88.

123 P. Dejardin, P. ten Hove, X. J. Xu and J. L. Brash, *Langmuir*, 1995, **11**, 4001.

124 C. F. Lu, A. Nadarajah and K. K. Chittur, *J. Colloid Interface Sci.*, 1994, **168**, 152.

125 P. Wojciechowski and J. L. Brash, *J. Biomater. Sci. Polym. Ed.*, 1991, **2**, 203.

126 L. Tang and J. W. Eaton, *Am. J. Clin. Pathol.*, 1995, **103**, 466.

127 W. G. Pitt, K. Park and S. L. Cooper, *J. Colloid Interface Sci.*, 1986, **111**, 343.

128 N. Bourdeau and M. Bissell, *Curr. Opin. Cell Biol.*, 1998, **10**, 641.
129 A. Huttenlocher, R. R.Sandborg and A. F. Horwitz, *Curr. Opin. Cell Biol.*, 1995, **7**, 697.
130 D. O. Schlaepfer and T. Hunter, *Trends Cell. Biol.*, 1998, **8**, 151.
131 E. Ruoslahti, *Annu. Rev. Cell Dev. Biol.*, 1996, **12**, 697.
132 S. Aota, M. Nomizu and K. M. Yamada, *J. Biol. Chem.*, 1994, **269**, 756.
133 M. Mrksich, *Chem. Soc. Rev.*, 2000, **29**, 267.
134 B. D. Bryan, T. W. Humment, D. D. Dean and Z. Swartz, *Biomaterials*, 1996, **17**, 137.
135 K. D. Chesmel, C. C. Clark, C. T. Brighton and J. Black, *J. Biomed. Mater. Res.*, 1995, **29**, 1101.
136 R. E. Baier, A. E. Meyer and C. K. Akers, *Biomaterials*, 1982, **3**, 241.
137 C. M. Stanford, J. C. Keller and M. Solursh, *J. Dent. Res.*, 1994, **73**, 1061.
138 P. J. Vezeau, G. F. Koorbusch, R. A. Draughn and J. C. Keller, *J. Oral Maxillofac. Surg.*, 1996, **54**, 738.
139 B. O. Aronsson, J. Lausmaa and B. Kasemo, *J. Biomed. Mater. Res.*, 1997, **35**, 49.
140 R. M. Shelton, A. C. Rasmussen and J. E. Davies, *Biomaterials*, 1998, **19**, 24.
141 K. Anselme, *Biomaterials*, 2000, **21**, 667.
142 C. H. Thomas, C. D. McFarland, M. L. Jenkins, A. Rezania, J. G. Steele and K. E. Healy, *J. Biomed. Mater. Res.*, 1997, **37**, 81.
143 C. R. Howlett, M. D. M. Evans, W. R. Walsh, G. Johnson and J. G. Steele, *Biomaterials*, 1994, **15**, 213.
144 G. Altankov and T. Groth, *J. Mater. Sci., Mater. Med.*, 1994, **2**, 732.
145 R. K. Sinha, F. Morris, S. A. Shah and R. S. Tuan, *Clin. Orthop. Rel. Res.*, 1994, **305**, 258.
146 D. A. Puleo, H. E. Preston, J. B. Shaffer and R. Bizios, *Biomaterials*, 1993, **14**, 111.
147 S. Ozawa and S. Kasugai, *Biomaterials*, 1996, **17**, 23.
148 G. Gronowicz and M. B. McCarthy, *J. Orthop. Res.*, 1996, **14**, 878.
149 J. Y. Martin, Z. Schwartz, T. W. Hummert *et al.*, *J. Biomed. Mater. Res.*, 1995, **29**, 389.
150 K. Kieswetter, Z. Schwartz, T. W. Hummert *et al.*, *J. Biomed. Mater. Res.*, 1996, **32**, 55.
151 B. D. Boyan, R. Batzer, K. Kieswetter *et al.*, *J. Biomed. Mater. Res.*, 1998, **39**, 77.
152 J. J. Qian and R. S. Bhatnager, *J. Biomed. Mater. Res.*, 1996, **31**, 545.
153 T. R. Pitt Ford, *The Restoration of Teeth*, Blackwell Scientific, Oxford, UK, 1985.
154 H. Ngo, G. J. Mount and M. C. R. B. Peters, *Quintessence Int.*, 1997, **28**, 63.
155 J. W. Nicholson, *Int. J. Adhesion Adhesives*, 1998, **18**, 229.
156 J. Vaughn, *The Physiology of Bone*, Clarendon Press, Oxford, UK, 1981.
157 P. Revell, *Pathology of Bone*, Springer-Velag, Berlin, 1986.
158 C. M. Stanford, J. A. Morcunde and R. A. Brand, *J. Orthop. Res.*, 1995, **13**, 664.
159 C. D. Toma, S. Ashkar, M. L. Gray, J. L. Shaffer and L. C. Gerstenfeld, *J. Bone Miner. Res.*, 1997, **12**, 1626.
160 Y. Mikuni-Takagaki, Y. Suzuki, T. Kawase and S. Saito, *Endocrinology*, 1996, **137**, 2028.
161 L. V. Harter, K. A. Hruska and R. L. Duncan, *Endocrinology*, 1995, **136**, 528.
162 J. Roelofsen, J. Klein-Nulend and E. H. Burger, *J. Biomech.*, 1995, **28**, 1493.
163 M. C. Maezzini, C. D. Toma, J. L. Shaffer, M. L. Gray and L. C. Gerstfeld, *J. Orthop. Res.*, 1998, **16**, 170.
164 U. Meyer, D. H. Szulczewski, K. Möller, H. Heide and D. B. Jones, *Cells Mater.*, 1993, **3**, 129.
165 P. Ngan, S. Saito, M. Saito, R. Lanese, J. Shanfield and Z. Davidovitch, *Arch. Oral Biol.*, 1990, **35**, 717.

166 A. J. Banes, M. Tsuzaki and J. Yamamoto, *Biochem. Cell Biol.*, 1995, **73**, 349.

167 L. L. Hench and J. Wilson, *An Introduction to Bioceramics*, ed. L. L. Hench and J. Wilson, World Scientific, Singapore, 1993, Chapter 1.

168 J. A. Lorenzo, S. L. Sousa, C. Alander, L. G. Raisz and C. A. Dinarello, *Endocrinology*, 1987, **121**, 1164.

169 P. E. Keeling, L. Rifas, S. A. Harris, D. S. Colvard, T. C. Spelsberg, W. A. Peck *et al.*, *J. Bone Miner. Res.*, 1991, **6**, 827.

170 T. Akatsu, N. Takahashi, N. Udagawa, K. Imamura, A. Yamaguchi, K. Sato *et al.*, *J. Bone Miner. Res.*, 1991, **6**, 183.

171 J. Pfeilschifter, C. Chenu, A. Bird, G. R. Mundy and G. D. Roodman, *J. Bone Miner. Res.*, 1989, **4**, 113.

172 S. C. Manolagas, *Ann. New York Acad. Sci.*, 1998, **840**, 194.

173 D. R. Bertolini and G. Strassmann, *Cytokine*, 1991, **3**, 421.

174 T. Albrektson, P.-I. Branemark, H.-A. Hansson and J. Lindstrom, *Acta Orthop. Scand.*, 1981, **52**, 155.

175 P.-I. Branemark, *J. Prosthet. Dent.*, 1983, **50**, 399.

176 P.-I. Branemark, U. Breine, R. Adell, B.-O. Hansson, J. Lindstrom and A. Ohisson, *Scand. J. Plast. Reconstr. Surg.*, 1969, **3**, 81.

177 L. Carlsson, T. Rostlund, B. Albrektsson, T. Albrektsson and P.-I. Branemark, *Acta Orthop. Scand.*, 1986, **57**, 285.

178 R. Adell, U. Lekholm, B. Rockler and P.-I Branemark, *Int. J. Oral Surg.*, 1981, **6**, 387.

179 R. Adell, U. Lekholm, B. Rockler, P.-I. Branemark, J. Lindhe, B. Eriksson and L. Sbordone, *Int. J. Oral Surg.*, 1986, **15**, 39.

180 T. Albrektsson, P.-I. Branemark, H.-A. Hansson, B. Kasemo, K. Larsson, I. Lundstrom, D. McQueen and R. Skalak, *Ann. Biomed. Eng.*, 1983, **11**, 1.

181 B. Kasemo, *J. Prosthet. Dent.*, 1983, **49**, 832.

182 G. R. Parr, L. K. Gardner and R. W. Toth, *J. Prosthet. Dent.*, 1985, **54**, 410.

183 C. M. Weiss, *J. Oral Implantol.*, 1986, **12**, 169.

184 T. Albrektsson, *J. Prosthet. Dent.*, 1983, **50**, 255.

185 T. Haraldson, *Scand. J. Plast. Reconstr. Surg.*, 1980, **14**, 209.

186 T. Albrektsson, G. Zarb, P. Worthington and A. Eriksson, *Int. J. Oral Maxillofac. Imp.*, 1986, **1**,11.

187 U. Lekholm, *J. Prosthet. Dent.*, 1983, **50**, 116.

188 D. B. Jones, S. B. Doty and R. C. van den Bos, *Failure in Joint Replacement*, ed. S. Downes, The University Press, London, 1993.

189 S. R. Golding, A. L. Schiller, M. Roelke, S. M. Rourke, D. A. O'Neill and W. H. Harris, *J. Bone Jt. Surg.*, 1983, **65A**, 575.

190 P. A. Revell, N. Al-Saffer and A. Kobayashi, *Proc. Inst. Mech. Eng.*, 1997, **211**, 187.

191 X. S. Zhang and P. A. Revell, *J. Mater. Sci., Mater. Med.*, 1999, **10**, 879.

192 J. C. Reed, *J. Cell Biol.*, 1994, **124**, 1.

193 E. White, *Gene Dev.*, 1996, **19**, 1.

194 D. A. Marlowe, *Biocompatibility Assessment of Medical Devices and Materials*, ed. J. H. Braybrook, John Wiley & Sons, Chichester, UK, 1997, Chapter 1, Part I.

195 M. Kuijpers and J. W. Dorpema, *Biocompatibility Assessment of Medical Devices and Materials*, ed. J. H. Braybrook, John Wiley & Sons, Chichester, UK, 1997, Chapter 1, Part II.

196 J. Tinkler, *Biocompatibility Assessment of Medical Devices and Materials*, ed. J. H. Braybrook, John Wiley & Sons, Chichester, UK, 1997, Chapter 10.

CHAPTER 7

Tissue Engineering

1 Introduction

Tissue engineering is a rapidly developing application of biomaterials that aims to create semi-synthetic body parts as alternatives to the use of donor organs.[1] The approach is characterised in that both artificial and natural biological materials are used, and the finished item is designed as a three-dimensional replacement for the appropriate body part. In reality, tissue engineering is a major field in its own right, and lies beyond the scope of this book. This chapter aims to provide a brief overview and readers are encouraged to consult the bibliography at the end of the chapter, as well as the individual references, for a fuller account.

Scope of Tissue Engineering

Typically tissue engineering involves an initial phase in which the device is developed by allowing cultured cells to grow *in vitro* onto and/or into a so-called scaffold which is usually made from a synthetic material, such as a polymer or a ceramic.[2] This allows the cells to proliferate freely, often aided by biological mediators such as growth factors. When this phase is complete, the entire device is implanted into or grafted onto the patient's body.

Tissue engineering has been driven by the needs of patients for substitute body parts, and this has been enhanced by the absence of sufficient numbers of fully natural tissues and organs for transplantation. It has been estimated that there is only one donor available for each patient in need of a transplanted organ.[3] Given this statistic, the potential for the emerging tissue engineering industry to make a dramatic impact on this problem is considerable.

In many cases, rather than a complete organ, it is a tissue that is required, as in some procedures in cardiovascular surgery. Heart disease is the major cause of death in the developed world, and there are approximately half a million coronary by-pass operations each year throughout the world.[4] About 25% of these patients lack the necessary native blood vessels for successful by-pass surgery. In addition, there are an estimated 100 000 patients who require replacement heart valves and other treatments for heart failure.[4] These add up to a considerable demand for heart and blood vessel substitutes, an area where, in the absence of sufficient numbers of donors, tissue engineering has the potential to make a major contribution.

As mentioned, tissue engineering may make use of synthetic materials, typically polymers or ceramics, as supports, so-called scaffolds. However, in many cases the scaffold is not artificial, but natural, such as collagen. The scaffold plays a crucial role in tissue engineering, since it not only supports the cells, but also guides cell adhesion, growth and the formation of new tissue in three dimensions.[5] In terms of structure, scaffolds are generally porous in order to allow cells to grow through the material, and mimic the natural tissue to be replaced. A number of such scaffolds have been developed from synthetic biodegradable polymers, such as poly(glycollic acid), poly(lactic acid) and glycollic–lactic acid copolymers,[6] and from hydrogels, such as poly(2-hydroxyethyl methacrylate)[7] or poly(vinyl alcohol).[8] In addition, natural polymer gels have also been used, including alginates.[9] Ceramics have also been used, most notably porous hydroxyapatite, as well as calcium phosphates and bioactive glasses.

Research is underway on various tissue engineered body parts, including bone, skin, heart valves,[10] nerves[11] and liver. The greatest success has been with skin, for which there are now commercial products based on tissue engineering. Progress is also good on tissue-engineered articular cartilage and on bone tissue, and these areas are now reviewed briefly.

2 Tissue Engineering of Skin

Skin substitutes are another area of high promise for tissue engineering, and where the approach is already making a positive impact on the treatment of patients. There is a need for such materials, for example for burns victims, and these people do not always have sufficient undamaged skin for skin grafting to be a viable treatment option. There is also a need for such materials to treat a variety of skin ulcers caused by systemic diseases, such as diabetes.

The first of the tissue engineered devices to be approved by the US Food and Drugs Administration was called Transcyte®. It is manufactured by a company called Advanced Tissue Sciences in La Jolla, California and consists of a single layer of skin cells, cultured from human neonatal foreskins supported on nylon mesh.[12] Another product of this type is Apligraft®, which is made by Organogenesis of Canton, Massachusetts and marketed by Novartis Pharmaceuticals. It is available commercially in the USA and Canada. It is also prepared from neonatal foreskins, but has a bi-layer structure (*i.e.* dermis and epidermis), which makes it closer to natural skin in morphology than Transcyte®. It has to be delivered fresh, and has a shelf life of 5 days.[13]

Tissue engineered skins have been shown to behave in an acceptable way, becoming readily incorporated into the recipient's body and gradually being replaced by fully functioning scar-free natural skin.[14] It has been used for a variety of purposes, including the treatment of diabetic foot ulcers,[15] chronic leg ulcers in patients with arterial and venous disease,[16] and minor burns.[17,18] Although very new products, tissue engineered skins are proving highly effective, results with them are good and their use seems likely to grow in the future.

3 Tissue Engineering of Articular Cartilage

Cartilage is the name given to the elastic tissue found in various parts of the body.[19] In the knee joint, acting as a lubricating component, the hyaline cartilage is found. This substance has a white, glassy appearance with no macroscopic evidence of fibres.[20] Both of the bones in the knee joints, the femur and the tibia, are covered with a layer of hyaline cartilage at the joint surface. The whole joint is encased within the synovial membrane and this forms the container for the synovial fluid. This synovial fluid provides the nutrients for the cartilage layers and is necessary because no blood vessels penetrate the cartilage from the subchondral bone.[21]

Hyaline cartilage is restricted to the knee, but two other kinds of cartilage are found elsewhere in the body, namely fibrocartilage and elastic cartilage. All types consist of chondrocytes and macromolecules of extracellular matrix, ECM. Fibrocartilage is found at the ends of tendons and ligaments in apposition to bone. It has a higher proportion of collagen in the ECM than hyaline cartilage.[22]

Cartilage may become damaged either through trauma or other type of injury. There are essentially three types of damage: matrix disruption, partial thickness defects and full thickness defects. Matrix disruption occurs from accidents, such as collisions with the dashboards in car accidents. The ECM becomes damaged but, unless the injury is serious, the remaining viable chondrocytes are able to repair the tissue.

Partial thickness defects occur through disruption of the cartilage surface, such as fissures, but this does not extend through to the subchondral bone. Although these defects are followed by proliferation of adjacent cells, this activity ceases before the defect can be repaired fully.

Finally, full thickness defects may occur. These arise from damage that extends through the entire thickness of the cartilage and penetrates to the subchondrial bone. In this case, the defect fills with a fibrin clot and a wound healing response follows. Unlike the other cartilage injuries, with this type of injury, progenitor cells from the bone marrow become involved in the healing process. They migrate to fill the defect[23] and cause replacement of the fibrin clot with a tissue that is intermediate between hyaline and fibrocartilage. This tissue is usually more permeable and of lower modulus than native cartilage, and this may contribute to its eventual degradation.

Overall, cartilage has very little capacity for self-repair following these injuries.[24] Chondrocytes, which might otherwise promote healing, are not required to proliferate to maintain cartilage in place, unlike other tissues such as skin. Consequently, one source of potential healing is absent. Secondly, except for full thickness defects, there is no direct access to proginator cells. Finally, the proteoglycans in the ECM can prevent cell adhesion, and this further undermines any potential healing process.[25]

One important approach to replacement of damaged cartilage is to use tissue engineering, with highly porous scaffold materials being employed to ensure the maintenance of cell differentiation in a given area. For chondrocytes, it is essential that they be grown in three dimensions if they are to maintain their differentiated phenotype and function.[26]

A variety of scaffold materials have been studied, both natural and synthetic,[27] but those based on collagen have received the most attention. The use of this material is beneficial because chondrocytes are able to maintain differentiated phenotype and production of GAG for six weeks when suspended in collagen gels.[28] Since this was discovered, there has been a considerable activity in employing collagen as a scaffold material, and promising results have been obtained. Collagen can be recognised by cellular enzymes and this means it can be remodelled to provide space for the developing tissue.[29] The use of collagen matrices also stimulates cells to produce new collagen.[30]

Although results for collagen, and also for derivatives of hyaluronic acid,[31] are promising, there are concerns about the use of a natural material. For example, there may be insufficient volume available for the required end use, and it may be difficult to ensure that all pathogens are removed. These concerns have stimulated research into the use of synthetic polymers as scaffold materials. To date, most interest in polymers has focused on degradable materials that have been approved by the US Food and Drugs Administration, *i.e.* poly(glycollic acid), PGA, poly(L-lactic acid), PLLA and their copolymer, poly(DL-lactic-co-glycollic acid), PGLA,[32–34] and research is continuing on this important topic.

4 Tissue Engineering of Bone

The aim of fabricating tissue engineered bone is to prepare filler materials for use in the reconstruction of large orthopaedic defects.[35] Traditional treatments include autografting and allografting. However, since they lack a vascular supply, the size of the defect that can be repaired is limited. There are other problems. In the case of autografts, operating time for collection material is expensive, and suitable bone is often in short supply.[36,37] On the other hand, the use of allografts carries with it the risk of disease and infection.[35]

Full bone regeneration requires four components, namely: (a) a morphogenetic signal, (b) host cells to respond to this signal, (c) a carrier for the signal and (d) a viable and well vascularised host bed.[38] Tissue engineered bone is fabricated to contain the first three of these, and needs to be placed in an appropriate location in order that there may be sufficient vascular tissue in a suitably healthy state.

In bone tissue engineering, a scaffold is used that either induces the formation of bone from surrounding tissue or acts as carrier for implanted bone cells. The materials may be injectable or rigid, and have included a variety of systems. These include calcium phosphate coatings on titanium plates,[39] hydroxyapatite–collagen composite[40] and porous hydroxyapatite.[41] The purely ceramic systems may be either absorbable or non-absorbable, and bio-absorbable polymers may also be used.[42]

An important feature, regardless of the type of material, is that the scaffold should be porous, though the actual shape of the pores required may vary depending on the function and location of the device. Optimum pore size has been shown to depend on cell type to be accommodated. For example, ectopic bone formation in ceramics is favoured by pore sizes in the range 300–400 μm,[43] whereas it seems to require a larger size range (200–400 μm) in natural hydroxyapatite derived from coral.[44]

Whatever biomaterial is used, the tissue engineered construct requires some sort of biochemical signalling molecule, typically bone morphogenetic proteins.[45–47] Their presence is essential in order to regulate the deposition of new bone and to promote healing in the traumatised bone.[48,49] Numerous experiments have shown that bone morphogenetic proteins will regulate the repair of defects in bone in a variety of animal models, including mice,[46] rats[50] and rhesus monkeys.[51] However, despite all the research activity and success in experimental animals, there is not yet a tissue engineered human bone repair system on the market. Research continues, though, and there is every prospect of such devices appearing in the foreseeable future.[35]

5 Bibliography

Further information on the broad subject of tissue engineering can be found in the following books:

R. Lanza, R. Langer and J. Vacanti, *Principles of Tissue Engineering*, 2nd Edition, Academic Press, New York, 2000.

J. Werkmeister, *Tissue and Cell Engineered Biomaterials and Medical Devices*, John Wiley & Sons, Chichester, UK, 2000.

F. Silver, *Biomaterials, Medical Devices and Tissue Engineering*, Kluwer Academic Press, Norwell, MA, 1993.

M. C. Shoichet and J. A. Hubbell (eds.), *Polymers for Tissue Engineering*, VSP International, Utrecht, The Netherlands, 1998.

C. W. Patrick Jr, A. G. Mikos and L. V. McIntyre (eds.), *Frontiers in Tissue Engineering*, Elsevier Science Publishers, New York, 1998.

6 References

1   R. Langer and J. Vacanti, *Tissue Eng Sci.*, 1993, **260** (5110), 920.
2   R. M. Nerem and A. Sambarris, *Tissue Eng.*, 1995, **1**, 3.
3   R. M. Nerem, *Proc. Inst. Mech. Eng., Part H*, 2000, **214**, 95.
4   R. M. Nerem, L. G. Braddon and D. Seliktar, *Tissue engineering and the cardiovascular system*, in *Frontiers in Tissue Engineering*, ed. C. W. Patrick Jr., A. G. Mikos and L. V. McIntyre, Elsevier Science Publishers, New York, 1998, pp. 561–579.

5   C. K. Kuo and P. X. Ma, *Biomaterials*, 2001, **22**, 511.

6   A. G. Mikos, A. J. Thorsen, L. A. Czerwonka, Y. Bao, R. Langer, D. N. Winslow and J. P. Vacanti, *Polymer*, 1994, **35**, 1068.

7   S. Lu and K. S. Anseth, *J. Control Rel.*, 1999, **57**, 291.

8   N. K. Mongia, K. S. Anseth and N. A. Peppass, *J. Biomater. Sci., Polym. Ed.*, 1996, **7**, 1055.

9   A. Martinsen, G. Skjak-Braek and O. Smidsrod, *Biotechnol. Bioeng.*, 1989, **33**, 79.

10 T. Shinoka, P. X. Ma, D. Shum-Tim, C. K. Breuer, R. A. Cusick, G. Zund, R. Langer, J. P. Vacanti and J. E. Mayer, *Circulation*, 1996, **94**, Suppl II, 164.

11 T. Hadlock, J. Elisseeff, R. Langer, J. Vacanti and M. A. Cheney, *Arch. Otolaryngol. Head Neck Surg.*, 1998, **124**, 1081.

12 G. Naughton, *Sci. Am.*, 1999, **280**, 84.

13 N. Parentu, *Sci. Am.*, 1999, **280**, 83.

14 N. L. Parenteau, J. Hardin Young and R. N. Ross, in *Principles of Tissue Engineering*, 2nd ed. ed. R. Lanza, R. Langer and J. Vacanti, Academic Press, New York, 2000, pp. 879–890.

15 M. L. Sabolinski and A. Veves, *Wounds*, 2000, **12** (Suppl A), 33A.

16 G. Bogensburger, W. Eaglstein and R. Kirsner, *Wounds*, 2000, **12**, 118.

17 J. Still and B. Craft-Coffman, *Wounds*, 2000, **12** (Suppl A), 58A.

18 D. W. Hayes Jr., G. E. Webb, V. J. Mandracchia and K. J. John, *Clin. Pediatr. Med. Surg.*, 2001, **18**, 179.

19 N. P. Cohen, R. J. Foster and V. C. Mow, *J. Orthop. Sports Phys. Ther.*, 1998, **28**, 203.

20 R. A. Stockwell, *Biology of Cartilage Cells*, Cambridge University Press, Cambridge, UK, 1979.

21 M. Harty, in *The Human Joint in Health and Disease*, ed. W. H. Simon, University of Pennsylvania, Philadelphia, PA, 1978, pp. 3–8.

22 J. S. Temenoff and A. G. Mikos, *Biomaterials*, 2000, **21**, 431.

23 J. A. Buckwalter, *J. Orthop. Sports Phys. Ther.*, 1998, **28**, 192.

24 E. B. Hunziker, *Osteoarthr. Cartil.*, 1999, **7**, 15.

25 E. B. Hunziker and E. Kapfinger, *J. Bone Jt. Surg.*, 1998, **80B**, 144.

26 A. M. Rodriguez and C. A. Vacanti, *Tissue engineering of cartilage*, in *Frontiers in Tissue Engineering*, ed. C. W. Patrick Jr., A. G. Mikos and L. V. McIntyre, Elsevier Science Publishers, New York, 1998, pp. 400–411.

27 C. J. Wirth and M. Rudert, *Arthroscopy*, 1996, **12**, 300.

28 T. Kimura, N. Yasui, S. Ohsawa and K. Ono, *Clin. Orthop. Rel. Res.*, 1984, **186**, 231.

29 D. P. Speer, M. Chrapil, R. G. Voltz and M. D. Holmes, *Clin. Orthop. Rel. Res.*, 1979, **144**, 326.

30 D. A. Grande, C. Halbertsradt, G. Naughton, R. Schwartz and R. Maryi, *J. Biomed. Mater. Res.*, 1997, **34**, 211.

31 L. A. Solchaga, J. E. Davies, V. M. Goldberg and A. I. Caplan, *J. Orthop. Res.*, 1999, **17**, 205.

32 C. A. Vacanti, W. Kim, B. Schloo, J. Upton and J. P. Vacanti, *Am. J. Sports Med.*, 1994, **22**, 485.

33 C. A. Vacanti and J. Upton, *Clin. Plast. Surg.*, 1994, **21**, 445.

34 C. R. Chu, R. D. Coutts, M. Yoshioka, F. L. Harwood, A. Z. Monosor and D. Amiel, *J. Biomed. Mater. Res.*, 1995, **29**, 1147.

35 K. J. L. Burg, S. Porter and J. F. Kellam, *Biomaterials*, 2000, **21**, 2347.

36 E. M. Younger and M. W. Chapman, *J. Orthop. Trauma*, 1989, **3**, 192.

37 B. N. Summers and S. M. Eisenstein, *J. Bone Jt. Surg. Br.*, 1989, **71B**, 677.
38 S. Croteau, F. Rauch, A. Silvestri and P. C. Handy, *Orthopaedics*, 1999, **22**, 686.
39 R. J. Dekker, J. D. De Bruijn, I. Van den Brink, Y. P. Bovell, P. Layrolle and C. A. van Blitterswijk, *J. Mater. Sci., Mater. Med.*, 1998, **9**, 859.
40 J. Ashina, M. Watanabe, N. Sakurai, M.Mori and S. Enomoto, *J. Med. Dent. Sci.*, 1997, **44**, 63.
41 J. A. Koempel, B. S. Patt, K, O'Grady, J. Wozney and D. M. Toriumi, *J. Biomed. Mater. Res.*, 1998, **41**, 359.
42 K. Whang, D. C. Tsai, E. K. Nam, M. Aitken, S. M. Sprague and P. K. Patel, *J. Biomed. Mater. Res.*, 1998, **42**, 491.
43 E. Tsuruga, H. Takita, H. Itoh, Y. Wakisaka and Y. Kuboki, *J. Biochem.*, 1997, **121**, 317.
44 R. E. Holmes, *Plast. Reconstr. Surg.*, 1979, **63**, 626.
45 T. Sakou, *Bone*, 1998, **22**, 591.
46 M. Suzawa, Y. Takeuchi, S. Fukumoto, S. Kato, N. Ueno and K. Miyazono, *Endocrinology*, 1999, 140, 2125.
47 T. Takahashi, T. Tominaga, N. Watanabe, A. T. Yokobori Jr., H. Sasada and T. Yoshimoto, *J. Neurosurg.*, 1999, **90**, 224.
48 M. P. G. Bostrom, J. M. Lane, W. S. Berberian, A. A. E. Missri, E. Tomin and A. Weiland, *J. Orthop. Res.*, 1995, 13, 357.
49 U. Ripamonti and N. Duneas, *Plast Reconstr. Surg.*, 1998, **101**, 227.
50 A. Linde and E. Hedner, *Calcif. Tissue Int.*, 1995, **56**, 549.
51 D. Ferguson, W. L. Davis, M. R. Urist, W. C. Hunt and E. P. Allen, *Clin. Orthop., Rel. Res.*, 1987, **219**, 215.

Subject Index